HOOVER'S

# Water Supply and Treatment

*Twelfth Edition*

by

Nicholas G. Pizzi

The Lake County Department of Utilities
Painesville, Ohio

BULLETIN 211
National Lime Association
Arlington, VA 22203

LIME "THE" VERSATILE CHEMICAL

KENDALL/HUNT PUBLISHING COMPANY
4050 Westmark Drive P.O. Box 1840 Dubuque, Iowa 52004-1840

Library of Congress Catalog Card Number: 94-78822

ISBN 0-8403-9625-2



Printed in the United States of America
10 9 8 7 6 5 4 3 2 1

New Dayton Plant

# In Memoriam

CHARLES POTTER HOOVER

1884–1950

Chemist—Engineer—Counselor—Humanitarian

*Member of American Water Works Association's Hall of Fame*

# PREFACE TO THE TWELFTH EDITION

The first edition, as well as the next six revisions, were written by Charles P. Hoover, who, in his forty-two years with the Columbus, Ohio, Water Department, established an international reputation as the Dean of Water Works Chemists.

The author of the eighth through the eleventh edition was Merrill L. Riehl, who was Superintendent of Purification and Chief Chemist of the Mahoning Valley Sanitary District (MVSD), Youngstown, Ohio until his retirement in 1978. I had the pleasure of meeting Mr. Riehl, a water works expert in his own right, at an AWWA function at MVSD in 1992. At that time, I had been working on the text for two years and was nearing completion of the project. It was a chance meeting - I had not expected to see him there. I spent the better part of the afternoon listening to him as he told me of his experiences as a student of Mr. Hoover, and of his subsequent years as Chemist for MVSD. I am thankful for the opportunity to have learned from him.

I wish to acknowledge the work on Chapters 9, 10, 11, and 13 which has been provided by Dr. Timothy A. Wolfe of Havens and Emerson, a Division of Montgomery Watson. We have endeavored to maintain the original flavor of the text, while attempting to introduce the latest treatment and regulatory concepts in this ever changing business.

Nicholas G. Pizzi
January, 1995

# ABOUT THE AUTHOR

Nicholas G. Pizzi is currently Director of Water Supply for the Lake County, Ohio, Department of Utilities. In addition he performs consulting services for municipal water districts in Ohio, and is a classroom instructor for both water treatment and water distribution systems. Mr. Pizzi received his bachelor's degree in Chemistry from St. Vincent College in Latrobe, Pennsylvania in May of 1971. He has more than twenty years experience in the water treatment field as a chemist, a plant superintendent, and an administrator. He holds a current Class IV Operator certificate in the State of Ohio. He has been active in AWWA as Ohio Northeast District Chairman, Chairman of the Ohio AWWA Education Committee, Chairman of the Community Relations Committee, as a member of several AWWA Technical committees, and as a PAC member for the AWWA Research Foundation (AWWARF). He has presented several papers at the Ohio AWWA annual conferences and has been published in the Journal, AWWA. Mr. Pizzi was a trustee on the Board of Directors for the Operator Training Committee of Ohio (OTCO) from 1988 to 1994.

# Contents

Settling Basins, Water Softening and Purification Works, Columbus, Ohio

Water Plant of the Mahoning Sanitary District, Youngstown, Ohio

Chapter I

# Source of Water Supply

When water falls upon the land it either flows into streams, lakes, and reservoirs, soaks into the ground, is absorbed by plants, or is evaporated. That which flows into streams, lakes, or reservoirs, or soaks into the ground can be made available for the public water supply. Water supplies are called surface supplies when they are taken from streams, lakes, or reservoirs, and ground supplies when taken from wells or springs.

Large municipalities make use of surface sources to secure an adequate supply of water, whereas smaller communities, private estates, and industrial plants ordinarily use underground sources. It is almost impossible to find a source of either surface or underground water that will meet the modern requirements for a public water supply without some form of treatment.

The requirements for a public water supply are that it meet all of the requirements of the Safe Drinking Water Act (SDWA), and that it be plentiful and as low in cost as possible.

## Surface Supplies

Surface water, whether from streams, lakes, or reservoirs, usually is contaminated and therefore unsafe and unsatisfactory for human consumption until properly treated. Municipalities sometimes discharge sewage into a water course that is used as a public water supply. This is perhaps the most dangerous source of contamination. Soil washings may carry mud, leaves, decayed vegetation, and human and animal refuse into the supply, thus rendering it turbid or unclean in appearance. The turbidity, or muddiness, and mineral content of water in flowing streams vary from day to day. Following heavy rains or freshets it may be extremely muddy and low in mineral content, whereas during dry seasons it may be relatively clear and more highly mineralized. Surface supplies may be muddy or clear, soft or hard, depending on the season. The immensity of the problem of mud removal from water taken from surface sources may be realized when we find that many of the large communities remove from 100 to 1000 or more tons of mud from a single day's supply of water before pumping it to the consumer. Water contaminated with human feces and urine may cause typhoid fever and dysentery. Organic wastes furnish food for microorganisms, which include vegetable growths such as algae and the lower forms of animal life. They may impart to the water disagreeable tastes and odors.

The waste liquors from manufacturing plants, mines, and quarries are often discharged into streams. These wastes may be objectionable because they are acid in nature, in which case they render the water either unfit for use or too corrosive for distribution through the ordinary iron or cement pipe distributing systems. Industrial wastes also may contain excessive quantities of organic material, which after decomposition causes the water to be unpalatable. Substances such as phenols from coke-oven plants are sometimes discharged into streams and lakes and are especially objectionable if chlorine is used in the treatment of the water supply. Chlorophenols are produced which impart to the water a disagreeable medicinal taste.

The watersheds of many streams are underlain with limestone and other rock formations containing soluble minerals. Water flowing through these deposits may dissolve compounds, causing the water to be highly mineralized and hard.

## Self Purification

Bacteria and algae consume and thrive on organic materials. Smaller microorganisms eat bacteria and algae, larger microorganisms live on the smaller ones, and fish and higher forms of animal life in turn consume the larger microorganisms. Unless the velocity of flow is too great, mud and organic matter is rendered harmless by oxidation, which is accomplished much more quickly and effectively if the stream flows over rough beds, riffles, and spillways. Sunlight has some germicidal effect due to its ultra-violet rays, but it penetrates only a little below the water surface and is not constant in its action because of night, and on some days, clouds block the sun's rays.

These natural processes, therefore, are not positive, and it is unsafe to generalize regarding the number of miles of flow necessary to accomplish self-purification. One river, for example, which received sewage for years, did not even become clear or lose its sewage odor until it had flowed some seventy or eighty miles. It was not bacterially safe even after many more miles of flow. Water in reservoirs after thirty or more days' storage may become practically free from turbidity and bacteria. However, this condition may be entirely changed in a few hours by the so-called overturning of the reservoir. This occurs when the surface water becomes chilled or warmed to a temperature of 39.2° F. At this temperature water has its greatest density. The heavier water, of course, cannot stay at the surface as convection currents are set up that may become sufficiently active to cause a complete overturn of the reservoir, resulting in much turbidity being diffused throughout the entire contents of the reservoir.

## Storage

In cities located on small streams which have a low flow during the hot or dry months, it is often necessary to impound water by means of dams and reservoirs in order to provide an adequate supply during all seasons. The volume of water in many streams varies from a raging torrent to almost no flow. The highest flood flow recorded in the Scioto River at Columbus, Ohio, if continuous throughout the year, would furnish more than enough water for all the people in the United States, yet during the summer and fall months the flow of this stream is not adequate to supply water for one-third of the present population of Columbus. Therefore, storm water must be stored in reservoirs in order to have an adequate supply during the low flow periods. For such storage a dam is usually constructed across a valley to form an impounding reservoir.

The ideal topographic conditions for a reservoir are a narrow gorge in which a dam may be built at minimum expense, and an expanding valley immediately above the gorge which will afford large storage per unit of surface area. The geology of the site frequently dictates the type of dam to be constructed. An earth dam may be suitable in a valley having deep impervious soil. Masonry construction is more desirable in a valley having solid rock walls and floor. Dams are made also of rock, wood, steel, and reinforced concrete, and are constructed after numerous patterns. The peculiarities of site, the money available, the magnitude of the project, and the use of the valley below the reservoir all influence the design of the dam.

## Reservoir Size

The necessary reservoir capacity is dependent upon the quantity of water required to supply the community, the amount of water lost by evaporation from the reservoir surface, by seepage through and around the dam, and upon the minimum flow of the stream. The minimum stream flow can be determined accurately only by actual measurement over an extended period of time. Even where such flow records are available there is always the possibility of a more severe drought than any previously recorded. However, all ordinary requirements will be met if provision is made for such droughts that occur not more often than once in 50 to 100 years.

The Water Resources Branch of the United States Geological Survey has measured the flow of many streams for years. Its publications (Water Supply Papers) contain many valuable data on observed stream flow. If the flow of the stream under consideration has not been measured, a study of the minimum yield of streams in the same territory, and of streams believed to be similar, will yield the best information obtainable.

In making such a study the varying physical conditions of the watershed should be noted. It has been observed that streams situated side by side may exhibit different run-off characteristics. Despite the difficulty of determining watershed similarities, results from comparison of streams similarly situated are usually more reliable than those obtained from run-off based on rainfall statistics. Even fragmentary records of the flow in the stream in question are valuable because when compared with the run-off records of nearby streams during the same periods the similarity of the watersheds is better determined.

To illustrate how the amount of storage required to carry a community through dry periods may be estimated, reference is made to Fig. 1-1, which shows a mass run-off curve for the Scioto River, above the O'Shaughnessy Dam at Columbus, Ohio, from March 1930 to February 1931, and the water demand rate for the city of Columbus during the same period, expressed in inches on the watershed above the dam. The demand rate may also include the estimated losses due to evaporation from the reservoir and to seepage around the dam. These have not been included in the one shown. Mass curves of run-off should cover years instead of months, but the illustration explains the principle. From the chart it will be noted that the total flow of the stream greatly exceeds the total requirements but that at intervals the slope of the run-off curve is flatter than that of the demand line. These flat places in the mass run-off curve represent periods during which the quantity of water passing down stream is insufficient to meet requirements. Storage is needed. The amount of storage required to carry the community through any of the recorded past droughts at the estimated rate of consumption may be determined by drawing tangents to the summits of the run-off curve parallel to the demand line, and measuring the greatest vertical distance between the two lines. Line AB is such a tangent. The arrowed dimension line gives the required storage.

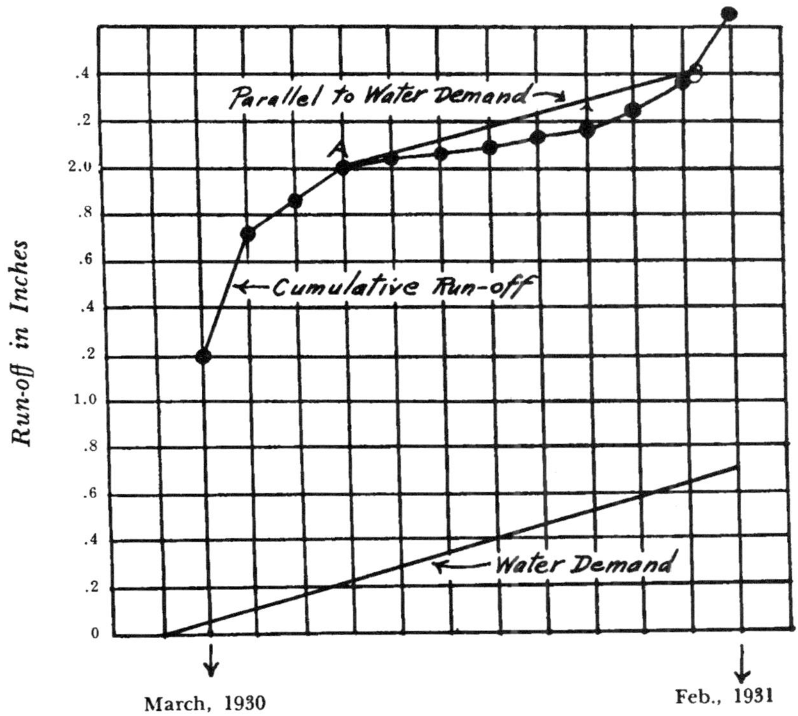


FIGURE 1-1. MASS RUN-OFF AND WATER DEMAND CURVES TO SHOW STORAGE REQUIREMENTS

Water from reservoirs or lakes should be examined at regular and frequent intervals in order to determine the approximate number of microorganisms present. If the results show the count to be increasing, treatment should be started before the growth becomes too profuse.

## Ground Water Supplies

Ground water is usually considered as being well water although it may include water from wells, springs, and infiltration galleries. Public supplies in this country are usually taken from wells. Well water supplies have certain advantages over surface supplies:

1. They are usually clearer.
2. They contain fewer bacteria.

3. If taken from a single well they usually have a uniform mineral content.
4. They usually have a more constant and lower temperature during the summer.

The temperature of ground water, coming from a depth of fifty feet, is the same as the average temperature of the region under which it lies. Water from a depth of less than fifty feet will be a little colder in the winter and a little warmer in summer. Water from a depth greater than fifty feet has a temperature higher than the average temperature of the region under which it lies. The temperature increases on an average of one degree for each sixty feet of depth below fifty. Some of the disadvantages of ground water are:

1. Its scarcity for large consumers.
2. Its uncertainty.
3. Calcium and Magnesium compounds are present in larger quantities than in surface waters found in the same localities.
4. Iron and manganese are present in many well supplies.
5. Hydrogen sulfide is often present.
6. The cost of pumping well water is usually greater than the cost of pumping surface water.
7. The mineral content from two wells may be entirely different even though located in the same plot of ground.

Calcium and magnesium compounds are objectionable in that they cause hardness; iron and manganese because they stain everything with which they come in contact; hydrogen sulfide by reason of its disagreeable taste and odor.

The extra cost of pumping well water depends on the depth from which it must be pumped and the type of pumping equipment used. Vertical centrifugal pumps have generally replaced air lifts and other types.

A public supply is often taken from a number of wells. If each well shows a different hardness content the softening process is complicated because the water from each source will require different quantities of softening chemicals.

Contrary to the opinion of many, ground water usually is quite limited in quantity. Larger cities almost always are dependent upon surface waters because not enough ground water may be obtained economically. Well supplies are particularly suited to the needs of smaller communities. In Ohio in 1992, there were 1277 community water supplies serving nearly 10 million people. Of these, 77% were well supplies, but surface water served 67% of the people. This example is typical. It shows that the total quantity of well water used by many small communities is greatly exceeded by the surface water used by only a few of the larger cities.

The yield of a well is dependent upon the porosity of the water-bearing formation, the size of the grains composing the soil or rock, the depth of the well into the water-bearing stratum, the rate of flow through this stratum, the diameter of the well, and the temperature of the water. If all these factors are known the yield can be calculated for any given draw-down, but since these factors usually are unknown, the yield of a well is best calculated by actual pumping. The water pumped is measured and the draw-down determined from time to time. The pumping should be continued until the water level in the well remains practically stationary. Continuous pumping for several days is frequently necessary to determine a reliable measure of the yield.

As long as the water level in the well is receding, water is being drawn from underground storage and for this reason the amount pumped is not a measure of continuous withdrawal. The water discharged during a few hours' pumping, after the water level becomes stationary, will furnish the best estimate possible as to the yield of a well. The yield of each well of a group generally is less than would be the case if the well were alone. The decrease in yield is due to mutual interference; that is, some of the water that would go to a well if it were alone is taken by others in the group.

The distance on either side of a well from which water is drawn is known as the radius of its circle of influence and, in general, it depends on the same physical conditions mentioned above as affecting the yield of a well. When wells are closer together than the diameter of the circle of influence, mutual interference takes place. Wells as far apart as 1000 feet usually have little or no effect upon one another with respect to yield.

It must be borne in mind that the action of ground water is not unlike that of surface water in that the water is flowing through the interstices of the water-bearing soil much as water flows in streams, but of course much more slowly. The velocity of ground water usually is only a few feet per day, even in coarse soil such as sand and gravel.

The amount of water that can be taken continuously from a well is dependent upon the quantity flowing through its circle of influence. Other nearby wells of similar size can increase the total quantity of water taken from the ground only by widening their circles of influence. The quantity derived from each well is, therefore, less than could be had by proper spacing of wells. A number of wells located in a line at right angles to the direction of underground flow will produce a greater yield than the same number of wells located in a line parallel to the direction of flow.

The diameter of a well has little effect upon its yield. Doubling the diameter of a well increases the flow by about 11 percent, and multiplying the diameter by 10 adds about 50 percent to the yield. The chief advantage of increased well diameter is that more room is provided for pumping equipment, and slower velocities through the well walls are obtained. The former makes possible more efficient pumps, the latter less trouble from entering sand and silt. Entrance velocities are improved by increasing the depth of penetration into the water-bearing stratum and increasing the screen depth accordingly, if screens are needed.

Where water is obtained from loose soils, sand, gravel, or glacial till, the materials from which the water comes should be analyzed for particle size so that the proper size of opening in the screen may be determined. The screen openings should be as large as possible, and yet exclude these loose materials.

## Quantity of Water Required

One of the early steps in the design of a water works system is the estimation of the quantity of water required. The estimate must of necessity be an approximation since it applies at some future date. However, a thorough understanding of all factors involved will greatly aid the estimator in arriving at a reliable answer.

The water consumption of a community usually is reported in gallons per capita per day. This quantity does not apply to any single day. It is the yearly average, and is based upon the total population of the community.

The chief factors tending to increase consumption, and which are more or less under the control of water works officials, are good quality, low rates, adequate service, and high pressure. The per capita water consumption in this country has been showing a gradual increase through past years due to the constant improvement of the quality of water offered the consumers, increased facilities for using the water, and greater ability to pay for it.

During droughts, when it was essential that the available water be conserved in an equitable manner, the responsible officials frequently have accomplished the desired end merely be reducing pressure, after other methods, such as newspaper appeals, have failed.

The nature and number of industries in a town greatly affect the water consumption. An industrial town, as a rule, has a greater consumption per inhabitant than one that is essentially residential. The demand of existing industries, as well as that of possible future industries, should be given careful consideration. The climate has its effect. A community in a region subject to hot, dry weather will require more water than one in a more temperate territory.

Probably the largest single factor affecting consumption is the presence or absence of meters. Where most of the services are on flat rates the consumption, (mostly waste of water), may reach a very high figure, in some cases more than 300 gallons per person per day. Since metering of public supplies is becoming more and more universal it is customary to assume that a large percentage of the taps will be metered when making quantity estimates. In fact, estimating flat rate consumption is almost futile.

For the purpose of estimating, it is frequently advantageous to divide the consumption into the various classes of user. The four classes usually employed are:

(1) Domestic; (2) Commercial and Industrial; (3) Public; (4) Loss.

## Water Consumption

Average figures based on reports from communities of 25,000 population and over:

| | *Gallons per capita per day (yearly average)* |
|---|---|
| Domestic | 64 |
| Commercial | 28 |
| Industrial | 37 |
| Public and other | 29 |
| Total | 149 |

These figures represent larger communities and will be less for smaller communities unless the services are not metered, in which case the consumption may be much higher.

## Variations from Annual Average

To properly design the various parts of a water works system it is necessary to know not only the average rate of water consumption, but also the variations from this average that may occur from time to time. Changing weather conditions and community activities cause a larger water use some months than others. Common habits of the people are such that on certain days of the week more water is used than on others. Likewise during some hours of the day the consumption is greater than during others. Knowledge of the relation that the monthly, daily, and hourly peak loads on a water supply system bear to the average load is of vital importance.

The hourly and daily variations from a 6-year average at the Lake County, Ohio, Aquarius Treatment Plant (20 MGD capacity) are exhibited in the following tabulation:

| | |
|---|---|
| Maximum quantity of water used in one day | 19.47 Million gallons |
| Minimum quantity of water used in one day | 8.58 Million gallons |
| Average quantity of water used per day (6 year period) | 11.3 Million gallons |
| Average day during 1991 | 11.4 Million gallons |
| Peak hourly flow for period | 24.3 MGD |

A large reservoir used for providing water through periods of drought should have its capacity predicated upon the monthly variation. Mains carrying water to the consumers need to be large enough to meet the maximum demands. Therefore, their sizes should be based upon the hourly variation. Pumps supplying direct pressure likewise must meet the maximum hourly demand, but if elevated storage is provided their size could be reduced to the requirements of the daily variation.

**Variations in Demand in Percent of the Annual Average**

| | *Large Cities* | *Small Cities* |
|---|---|---|
| Monthly | 110 to 130 | 120 to 150 |
| Daily | 125 to 175 | 150 to 250 |
| Hourly | 230 to 300 | 300 to 1000 |

The foregoing table illustrates the range of variations above the annual average most commonly found in existing supplies. The dividing line between large and small cities is vague, but may be taken at about 50,000 population.

## Fire Demand

Although the total quantity of water used annually for fire fighting is a very small part of the total consumption, the rate at which this water is needed is often an important factor in water works design, particularly within small cities and towns.

### *Fire Demand Rate*

**Required Flow for Principal Business Districts of Average Cities**

| *Population* | *GPM* | *MGD* | *Hours duration* |
|---|---|---|---|
| 1,000 | 1,000 | 1.44 | 4 |
| 1,500 | 1,250 | 1.80 | 5 |
| 2,000 | 1,500 | 2.16 | 6 |
| 3,000 | 1,750 | 2.52 | 7 |
| 4,000 | 2,000 | 2.88 | 8 |
| 5,000 | 2,250 | 3.24 | 9 |
| 6,000 | 2,500 | 3.60 | 10 |
| 10,000 | 3,000 | 4.32 | 10 |
| 13,000 | 3,500 | 5.04 | 10 |
| 17,000 | 4,000 | 5.76 | 10 |
| 22,000 | 4,500 | 6.48 | 10 |
| 27,000 | 5,000 | 7.20 | 10 |
| 33,000 | 5,500 | 7.92 | 10 |
| 40,000 | 6,000 | 8.64 | 10 |
| 55,000 | 7,000 | 10.08 | 10 |
| 75,000 | 8,000 | 12.96 | 10 |
| 95,000 | 9,000 | 12.96 | 10 |
| 120,000 | 10,000 | 14.40 | 10 |
| 150,000 | 11,000 | 15.84 | 10 |
| 200,000 | 12,000 | 17.28 | 10 |

Over 200,000 population, 12,000 gpm, with 2,000 to 8,000 gpm additional for a second fire, for a 10-hour duration.

The foregoing fire rates are the recommendation of the National Board of Fire Underwriters. This board also recommends that these rates be maintained at a minimum pressure of 20 pounds per square inch, except in certain specified conditions where a minimum of 10 pounds per square inch is permissible.

## Future Demand Based on Population Forecast

Most water works structures should be built larger and have a greater capacity than is required for immediate needs. How far future needs should be anticipated depends upon the nature and permanence of the structures, and upon the cost of capital. An equitable division of the financial burden between the present and future consumers usually dictates that provision be made for 10 to 30 years in advance of present requirements.

The various methods for predicting the growth of towns and cities are all based either upon the past growth of the city under consideration, or upon the growth of similar and larger cities. Methods dependent upon past growth are: (1) graphic projection; (2) arithmetical increase; (3) percentage increase; (4) decreasing percentage increase.

The graphic projection is made by plotting populations vertically and years horizontally on a coordinate chart, and then drawing through the points a smooth curve projected into the future "by eye." The knowledge and bias of the person making the projection will be reflected in the estimate obtained.

An arithmetical forecast is made by adding to the present population the product of a fixed sum multiplied by the number of future years it is desired to include. The fixed sum is the annual growth during a few past years. The method generally gives results too small except for those cities which are limited in their territorial growth, and where population increase is dependent almost entirely upon increased density.

By percentage increase method the percentage of increase in population during the previous decade, or the average of several past decades, is determined. This percentage is assumed to be constant for all future decades and is applied to each succeeding decade in turn. Results are usually too large except for rapidly growing cities.

A modification of the preceding method makes use of a decreasing percentage of increase in population each succeeding decade. The amount that the percentage will be decreased each decade will depend upon the previous rate of decrease as revealed by census statistics of the municipality in question and upon the judgment of the forecaster.

The growth of the city in question is plotted upon a chart, with years as abscissas and population as ordinates. The growth curves of several larger cities, preferably those having similar geographical and commercial characteristics, are plotted on the same chart in such a way that they all pass the present population of the city under consideration. With the curves from these other cities as guides the population curve of the city in question is then projected into the future.

Obviously, any forecast which is to be more than a rough approximation must be preceded by a careful analysis of all the factors affecting the community as well as long time records and trends of the particular community and of similar communities. In any event the accuracy of such forecast will in a large measure depend upon the skill and experience of the observer.

## Chapter II

# Methods of Water Purification

In general, there are seven well-known methods of water purification; namely, prolonged storage; filtration through slow sand filters; coagulation and filtration through rapid sand filters; combined water softening and iron removal, coagulation, and filtration through rapid sand filters; water softening; iron and manganese removal; and disinfection.

In addition to these, there are a number of special treatment processes and modifications. One of the special treatment methods now receiving much attention is desalination, a process which may be accomplished by various methods, and which is primarily used where only saline or brackish water is available. Other special processes, which also will be described briefly, are "straining" by new patented devices and diatomaceous earth filtration. Very often it is necessary that a combination of treatments be used in order to obtain desired results.

This treatise does not cover water plant design. For information on this subject the 1990 text, "Water Treatment Plant Design," copyrighted by the McGraw-Hill Company. New York, NY., is recommended.

## Purification by Storage

Purification by storage involves storing water in basins or reservoirs for long periods of time. This affords an opportunity for the subsidence of silt and clay, reduction in color, and the death of bacteria. Considerable time is required for these results to be accomplished by storage alone, and even then bacteria and finely divided particles of clay are not entirely removed. The sizes of the basins or reservoirs required make this method very expensive. The tendency toward stagnation and the multiplication of low forms of animal and vegetable growth make it undesirable. Even though many days of storage are provided, other methods of treatment usually are required to insure a satisfactory product. In many cases additional treatment for algae control supplements the natural purification steps. Stratification also is avoided in some cases by mechanical devices for aeration.

## Slow Sand Filtration

Slow sand filters, while still used rather extensively in Europe, are no longer being built in this country. A few older plants may still use this type of filter, but most have been either supplanted or their use supplemented by additional pre- or post-treatment.

The main disadvantages of slow sand filters are:

1. The large area required for construction of filters.
2. Their successful operation is limited to clear water with low color.
3. The first cost is relatively high.

## Straining

A device utilizing extremely fine-mesh screens has been developed for use where raw supplies are rich in algae and plankton. They are used as the sole filtering process, or in combination with granular filters. The "strainers" are usually a large horizontal cylinder, open on one end, with the fine screen held in place around the circumference.

PICTURE 2-1. WATER SOFTENING PLANT, MINNEAPOLIS, ILLUSTRATING SLUDGE BLANKET TYPE WATER SOFTENER

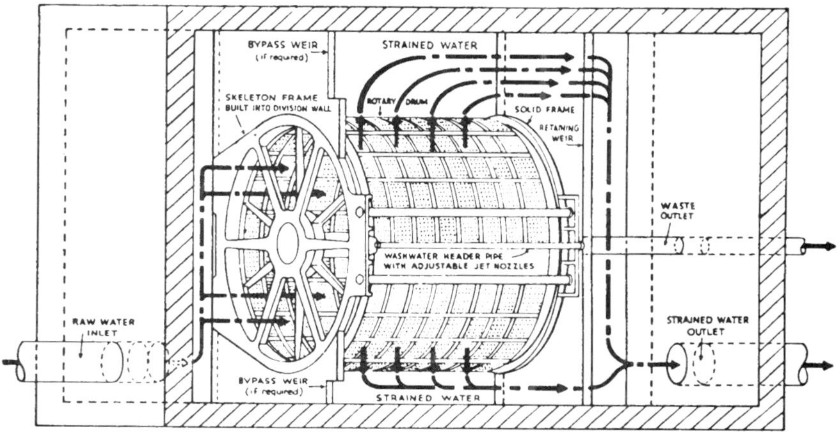


FIGURE 2-1. PERSPECTIVE VIEW OF GLENFIELD "MICROSTRAINER"

The raw water flows in the open end and by a slight gravity head loss is forced through the fine mesh, removing most plankton and sediment. The rotating drum is only partly submerged, and the exposed part continuously subjected to flushing water which washes the collected solids into a trough and to waste. Rates of 5 to 25 gpm/ft$^2$ of submerged area are reportedly used in water supply operations.

As a rule these units, whether as a pretreatment or the only treatment (besides disinfection), are limited to waters low in turbidity and color. See Figs 2-1 and 2-2.

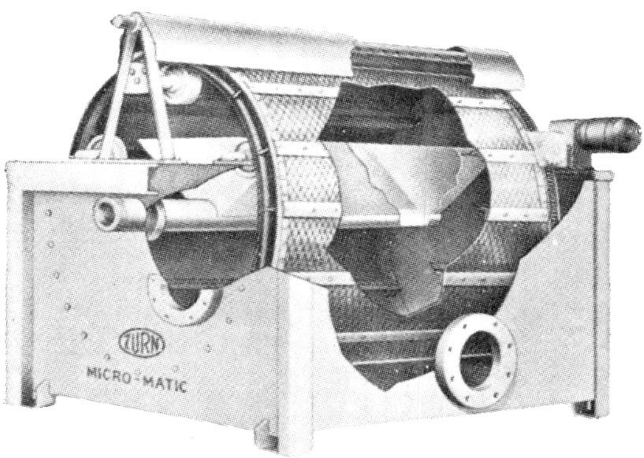


FIGURE 2-2. ZURN "MICROMATIC STRAINING SYSTEM"

PICTURE 2-2. FILTER GALLERY, BENTON HARBOR, MICH.

## Rapid Sand Filtration

To obtain a high rate of filtration through sand filters (100 to 200) million gallons per acre per day), it is usually necessary to treat the water ahead of the filters. This preparatory treatment includes: (1) aeration to free the water from dissolved gases and to oxidize iron and organic matter if present; (2) pre-chlorination; (3) coagulation; and (4) settling.

The addition of the coagulant to the water, followed by a short period of agitation, results in the formation of a precipitate which entangles the mud, bacteria, and suspended matter into clumps, readily removed by settling. The settled water then passes to the filter which removes the finely divided suspended matter that does not settle in the settling tanks.

Water, when taken from a flowing stream, is usually passed through coarse screens into a suction well. The screens exclude fish, sticks, leaves, and other floating material. Low lift pumps force the water through a meter to the filtration plant. (See Fig. 2-3) It enters a rapid mix tank, usually at the bottom where a coagulating chemical is added. The coagulant is stirred into the water, usually for about 5–45 seconds. Rapid mixing is usually followed by slow mixing, or flocculation, which can last from 15–45 minutes or more. From the overflow at the top of

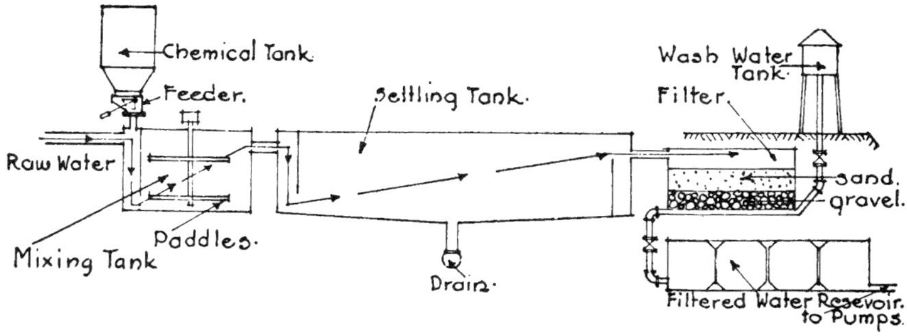


FIGURE 2-3. FLOW CHART OF RAPID WATER FILTRATION PLANT

the flocculator the water flows to the settling basin where the insoluble compounds or precipitates resulting from the coagulation and flocculation treatment settle to the bottom. From the top of the settling basin the partially clarified water flows to the filters where it passes through sand, or other suitable filtering material, into a storage basin or reservoir. On its way to the storage basin a small amount of chlorine is usually added. Rapid sand filtration will be explained more fully in later chapters.

## Aeration

Many water supplies do not need to be aerated. Surface supplies, however, sometimes have offensive tastes and odors. Aeration may remove them if they are due to dissolved gases resulting from the decomposition of organic matter. If, however, they are due to dissolved organic matter, aeration will not be very effective.

Aeration covers both physical removal of undesirable gases as well as chemical removal by precipitation of dissolved minerals such as iron and, to some extent, manganese.

Ground supplies usually contain carbon dioxide, and sometimes hydrogen sulfide, iron and manganese. Aeration reduces the carbon dioxide and hydrogen sulfide and, in most cases, oxidizes the iron, causing it to precipitate as an insoluble ferric compound which can be removed readily by filtration. Manganese, especially if it is present in appreciable quantities, is not so easily removed as iron. Aeration and filtration usually are not adequate. Iron and manganese can be precipitated from aerated water very effectively by the addition of lime. Manganese requires more lime for its precipitation than iron. A pH of

PICTURE 2-3. SHOWING AERATING EQUIPMENT PLACED OVER EXISTING BASIN, MONROE, LOUISIANA LIME AND IRON USED PRIOR TO FILTRATION.

9.4 is usually adequate for the precipitation of manganese. The reduction of carbon dioxide by aeration decreases the cost of softening water if lime is used as the reagent. One objection to the aeration of oxygen-free ground water is that it becomes oxygenated and under some circumstances may be more corrosive.

Feeding of chlorine occasionally is used as a substitute for aeration; particularly for removal of hydrogen sulfide, iron, or manganese. Where a change in raw water necessitates such treatment after a plant is built or only during intermittent periods, addition of aeration devices may be more difficult than installation of chlorine-feed equipment.

Aeration is accomplished by various means. It may be done by allowing the water to flow over cascades or steps, by spraying from spray nozzles, by trickling through trays filled with coke, or by wood slot and forced draft aerators, or by other means. Photographs on pages 20, 109, and 190 show several aerators in use; also figs. 2-5, 2-6. and 2-7.

## Desalination

Lack of fresh water has retarded the development of some areas of the world. Until suitable methods of treating the saline of brackish water became available, transportation of fresh water over great distances by rail, truck, ship, or pipeline was the only alternative and was too costly for any but extremely meager use.

In the last three decades, intensive research has been carried out to utilize waters which were formerly considered too high in solids to be of any value. Many different procedures have been tried to remove or reduce the salt content of these waters eventually resulting in techniques which, while expensive, made water available at a price which was economically feasible for the purposes intended. Much more research is still being carried out; it is expected that more efficient methods will be developed which will further reduce the processing costs.

Some of the methods* used, or under study, are:

1. Multi-stage flash distillation
2. Multi-effect evaporation
3. Electrodialysis
4. Reverse osmosis
5. Vacuum freezing
6. Secondary refrigerant process
7. Solar distillation
8. Ion exchange system
9. Hydrate process
10. Vapor reheat flash distillation

Some procedures are more adaptable to brackish waters (up to 3000 mg/L total solids) than to saline waters (up to 35,000 mg/L total solids). Also some of the methods produce essentially solids-free water, while others reduce the solids to a level acceptable for the use intended.

Selection of the processes is a complex one involving the type of water available, the degree of treatment required, the quantity of water needed, fuels available, availability of waste heat from electric generating plants or other industry, and the demand, i.e., continuous or variable.

The multiflash evaporation method is probably the method producing the largest quantity of treated water; however, reverse osmosis and electrodialysis appear to have merit for future development. In particular, the reverse osmosis process is finding acceptance as a viable "point of use" treatment technique for the removal of inorganic salts and heavy metals. Quite a number of desalination plants have been built

---

*From "Desalination and Its Role in Water Supply", U.K. Atomic Energy Auth., 1967

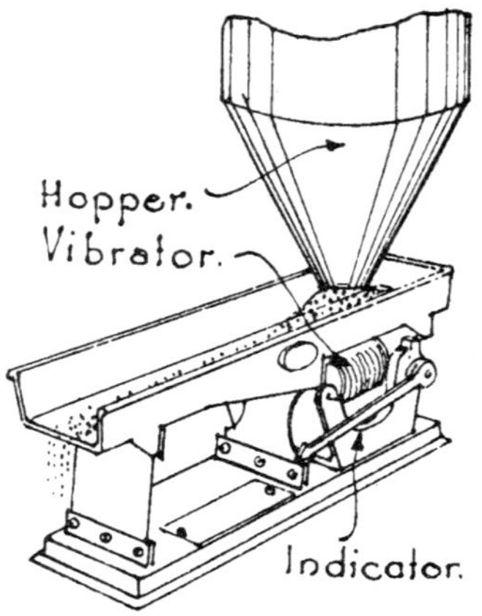


In this type feeder the dry chemical drops from the hopper onto the deck which vibrates about 7200 times per minute. The quantity shaken off the end of the deck depends on the intensity of the vibration, which can be regulated as desired by raising or lowering the indicator arm.

FIGURE 2-4. CHEMICAL FEEDER VIBRATOR TYPE

and put into operation. Many more are expected as the world economic situation makes it necessary for industrial development of many new areas.

## Chemical Treatment and Application

The addition of chemicals to water is usually the first step in the treatment process. The chemicals may be applied in a dry pulverized condition or they may be put into solution or suspension and fed into the water by means of a measuring device. Liquids of predetermined strength are prepared by dissolving or suspending a weighed quantity of chemical into a specified volume of water. Dry feed machines equipped with continuous dissolvers or slakers have been developed and are coming into general use. They automatically measure and weigh the pulverized dry chemical.

Dry feed machines are of two types; one measures the chemical by volume and the other by weight. There are many modifications of these devices but those illustrated show, in a general way, how each works.

The quantity of chemicals required to condition water for filtration usually is very small. It may vary in amount from a maximum of 1000 pounds to a minimum of less than 100 pounds per million gallons. Unless the water plant is a large one treating millions of gallons of water daily, or a softening plant requiring large quantities of chemicals, labor-saving devices are not generally needed. However, softening for even a relatively small community may justify the installation of rather elaborate chemical handling devices.

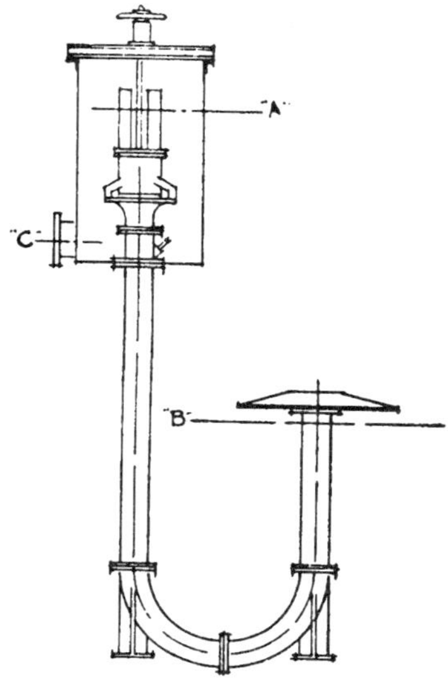


**AER-O-MIX**

Water enters at (C), rises to level (A), flows down draft tube and discharges over splash plate (B). The head consists of hollow tubes open to outside air above (A). Water passing around them aspirates air into the down-flowing water which is held until released at splash plate.

**SPLASH AERATOR**

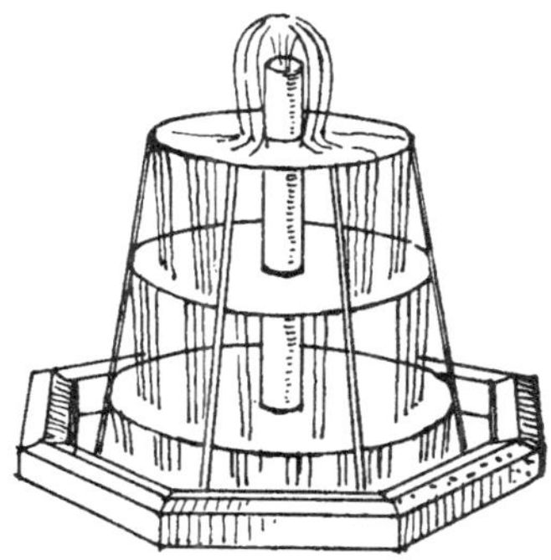

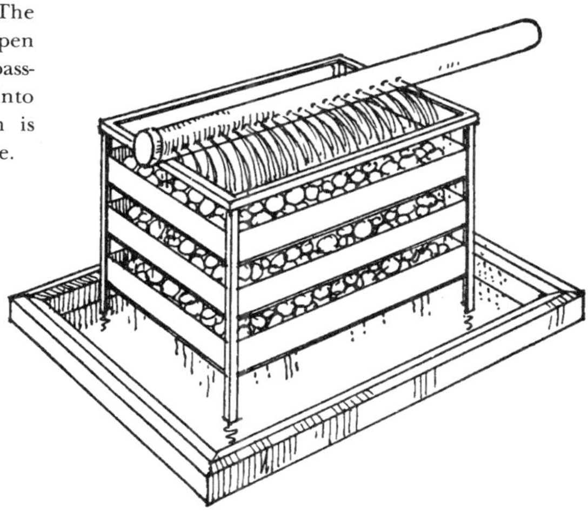

**COKE TRAY AERATOR**

Rate of filtration usually is not in excess of 10 gallons per square foot per minute.

FIGURE 2-5. TYPES OF AERATORS

Methods for handling, storing, and applying large quantities of chemicals were in the past very crude but now large quantities can be handled economically. Power shovels and pneumatic conveying devices have come into use. Elevation into bins from which the chemicals subsequently are fed into the water by gravity costs little. An explanation of the operation of the pneumatic system now being generally adopted is shown in Fig. 2-10.

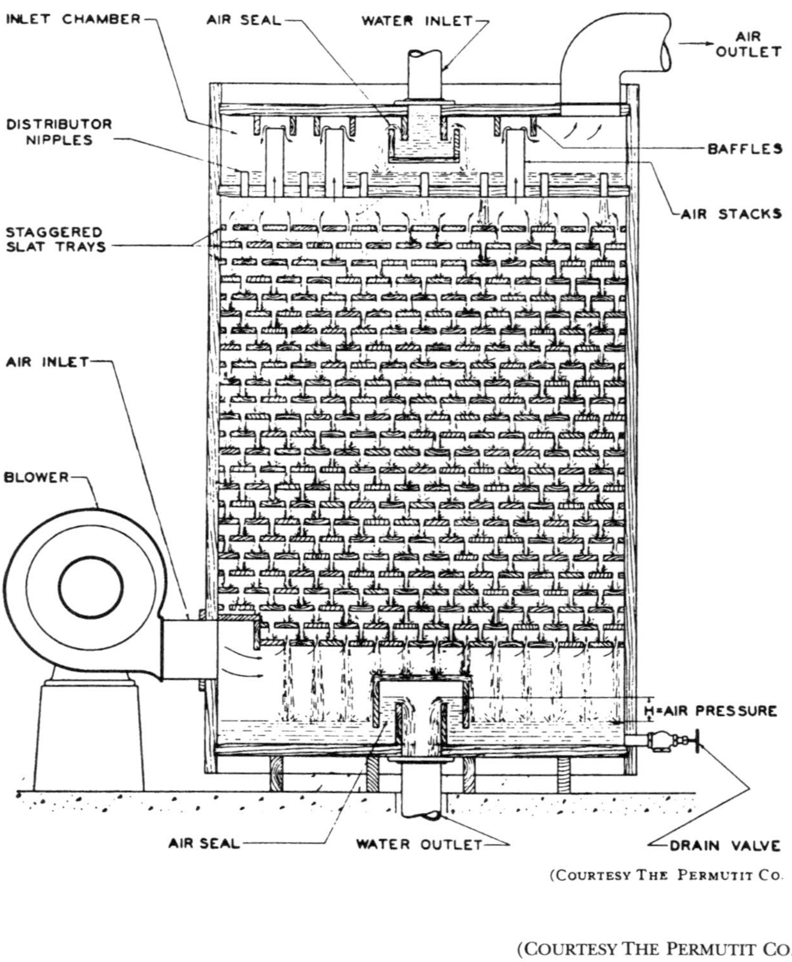


(COURTESY THE PERMUTIT CO.)

FIGURE 2-6. DEGASIFIER OR FORCE DRAFT AERATOR

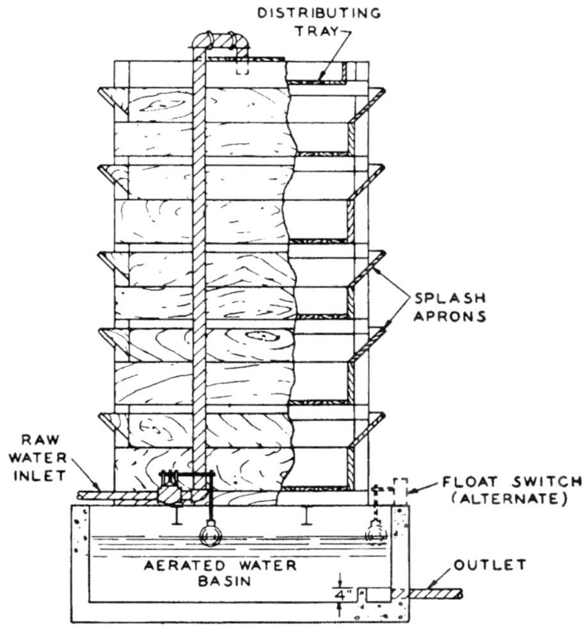


FIGURE 2-7. WOOD SLAT TRAY AERATOR

With the trend to delivery by trucks, which carry their own source of compressed air, there is a rapid trend to positive pressure unloading systems for bulk dry chemicals. See Fig. 2-11. Such a system exists at the Bureau of Reclamation Yuma Projects Office in Yuma, Arizona. This 100 MGD water treatment facility is capable of storing 6000 Tons of Lime in its silos. The plant was designed to use 100 Tons of Lime per day. When dry chemical strorage is not possible or recommended, liquid chemicals are used so tank delivery direct to solution tanks from trucks is common practice.

Combining the two, many plants now receive the dry chemicals in bulk but discharge from trucks into tanks where water is added to form a slurry in which condition the chemicals are stored and used. This is particularly true of carbon but is coming into consideration for soda ash.

It is very important that the application of chemicals to water be carefully regulated and controlled in order to prevent overtreatment or undertreatment. The application of chemicals, especially when softening, should be proportional to the volume of water coming to the plant. Unless the flow is constant, or approximately so, feed mecha-

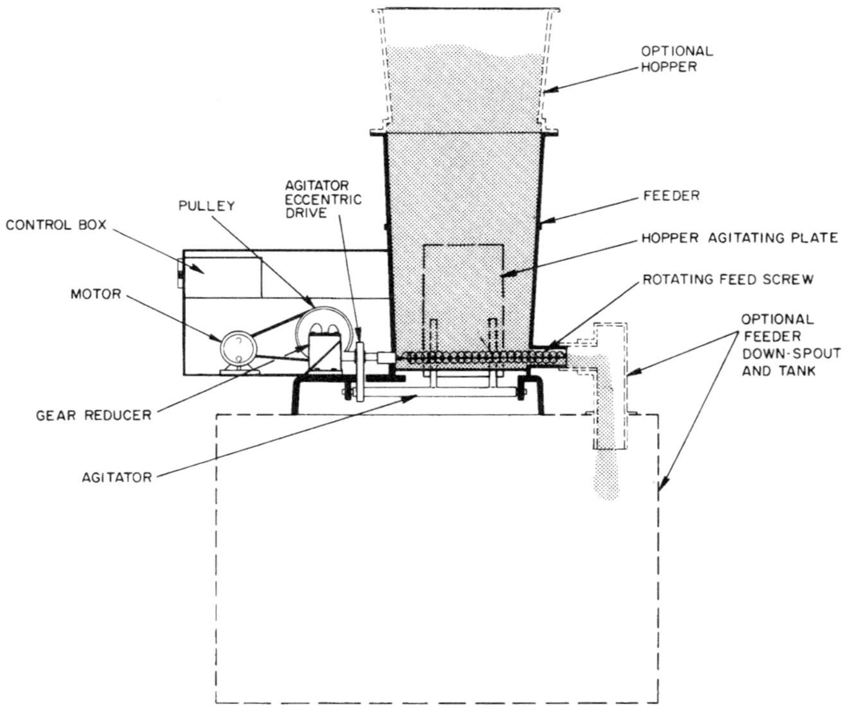


How Scale Operates

1. Poise is set to balance weight of chemical in hopper.
2. Poise is driven inward on beam at desired rate.
3. Inward movement of poise unbalances scale beam completing electrical circuit which starts feeder.
4. Discharge of chemical from hopper causes scale beam to gain balance, thus breaking contact and stopping feeder.
5. This cycle continues until hopper is empty.

FIGURE 2-8. CHEMICAL FEEDING DEVICE FOR CONTINUOUSLY WEIGHING AND DISSOLVING CHEMICALS

nisms should be provided with devices for automatically increasing or decreasing the application of chemicals in direct proportion to the increase or decrease of the flow. This is accomplished by various methods. Dry feed weighing machines may be electrically controlled through a meter that makes an electrical contact for every unit volume of water (such as 500, 1000, 2000 or more gallons) passing through it. Such contact closes a circuit which starts the feeder operating. The duration of the contact is regulated so that the desired dose for that quantity of water is fed, then the feeding machine stops until contacted again by the meter, thus continuing the cycle. Other types of dry feeders are controlled by mechanical means actuated by a meter to change the speed or number of cycles of the feeder in proportion to the increase or decrease in the flow.

Solution machines have, in general, two types of control. In one, an orifice in the end of a swinging draw-off pipe is at the surface of the solution in the tank. Attached to this pipe is a wire or cable, the other end of which is wound on a drum attached to a meter. As the meter shaft revolves with the flow of the water the drum unwinds, lowering the end of the pipe into the liquid. This permits a controlled quantity of the solution to be fed into the raw water.

The pneumatic unloading system consists of an intake pipeline, including flexible hose and intake nozzle, receiving station, the filters for cleaning the conveying air, and the vacuum pump. To unload a car of chemicals, the vacuum pump and motor to drive the discharge

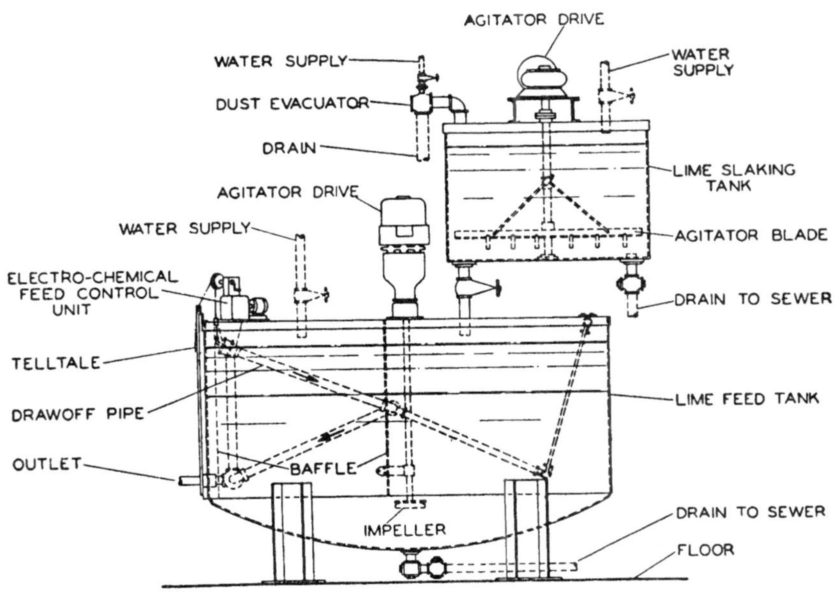


FIGURE 2-9. WET TYPE CHEMICAL FEED WITH AUXILIARY LIME SLAKING TANK

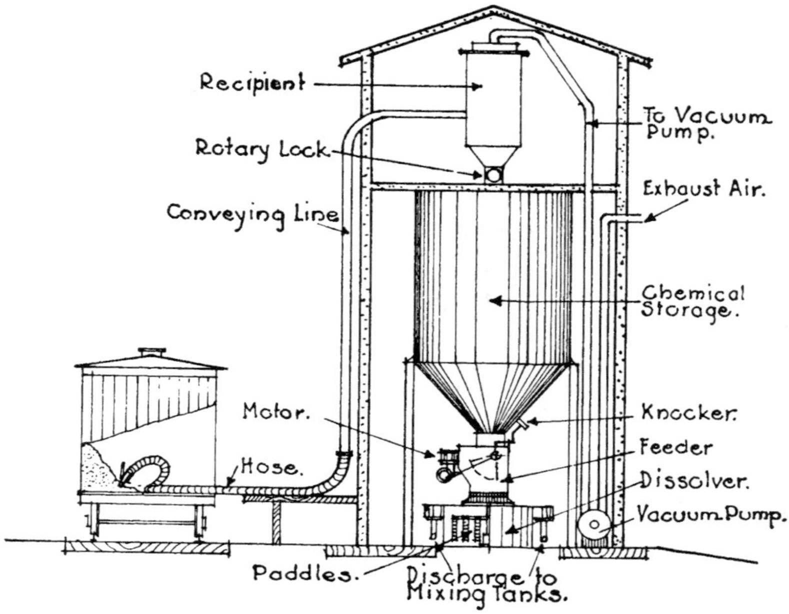


FIGURE 2-10. PNEUMATIC UNLOADING, STORAGE AND FEEDING OF CHEMICALS

mechanism under the receiving station are started. Atmospheric air is drawn in at the intake nozzle, and when the nozzle is immersed in the material, the inflow of air picks up the chemical and conveys it in a steady stream at high velocity to the receiving station. There the abrupt decrease in velocity causes the precipitation of the material in the hopper, when the conveying air is drawn up through the filters, which consist of three cylindrical housings containing cloth tubes. The air passes up through the tubes, but the cloth retains the fine dust particles, so that only clean air is drawn through the pump and discharged to the atmosphere. There is an automatic arrangement to clean the dust regularly from the cloth tubes in each filter compartment, so the dust feeds out of the system continuously with the coarser particles. The chemical is discharged through the rotary lock in the hopper, from which it is measured or weighted into the slaker or dissolver and thence into the water being treated.

The other type of control makes use of the differential in pressures between the throat and the mouth of a Venturi meter. This differential

PICTURE 2-4. BUREAU OF RECLAMATION—YUMA, ARIZONA—LIME SILOS.

pressure is utilized in connection with floats and levers to regulate the position of a cut-off shield or gate over the out pipe from the chemical solution tank.

Other types of feeders for liquids also are paced to feed the chemical in proportion to the flow. Some of those commonly used are the BIF "Rotodip" and the small diaphragm-type feeders furnished by various manufacturers.

## Lime Slaking and Feeding

The slaking qualities of lime depend upon the size of lumps to be slaked and also on how the lime was burned.* So-called pebble or rotary kiln lime usually is made from washed, small stones and is burned very rapidly. Fast burning results in quick expulsion of the carbon dioxide from the stone and the formation of a very porous product. When water is added to highly porous lime, it slakes very quickly because of the large surface area exposed to the slaking water. For example, when 100 grams of fresh quicklime are placed in 400 grams of water at a temperature of 25°C. (77° F.) and stirred, the temperature should rise to at least 72°C. (or 162° F.) in 3 minutes, whereas with slow-slaking lime there is very little rise in temperature in 3 minutes.

When quicklime is fed by a dry feeder into a continuous slaker it usually is necessary to specify quick-slaking or finely ground lime. Four to five pounds of water to one pound of quicklime generally are sufficient for good slaking. Too little water produces a slurry that is too thick to be agitated properly, and too much water not only chills the material being slaked but reduces the retention time in the slaker. The quantity of water should be regulated so as to keep the temperature of the water in the slaker at about 66° to 77°C. (or 150° to 170° F.). When slaking temperatures are too high the operating room steams excessively if the slaker is not provided with a vapor remover. When the slaker is provided with a vapor remover the steam and vapor overload it and the escaping moisture causes corrosion of the equipment. When quicklime is slaked in an auxiliary lime slaker, then discharged into a lime feed tank and measured as a milk of lime into the water being treated, it usually is diluted. It is possible to feed solutions having as low as four pounds of water to one pound of lime but more dilute solutions have the advantage in that they can be applied more accurately. The general practice is to dilute to a 10 per cent solution.

Of comparative recent development are slakers which need a much smaller amount of water, leaving the hydrated lime in the form of a

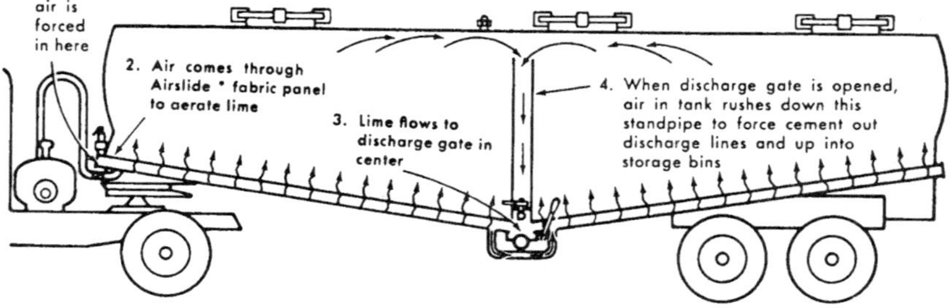


FIGURE 2-11. POSITIVE PRESSURE SYSTEM FOR UNLOADING BULK LIME FROM TRUCK TRAILER

*For more detailed information on lime, see Boynton, R.S., "Chemistry and Technology of Lime and Limestone," John Wiley & Sons, Inc., New York (1980).

PICTURE 2-5. DRY LIME FEED MACHINE, WHEELING, W. VA.

very thick slurry or "paste". These units have been found to be quite efficient, and they are being used rather extensively (See illustration)

With the solution type of feeder it is essential to have thorough agitation in order to prevent the lime suspension from stratifying. Stratification prevents uniform application of the lime to the water, thereby producing irregular results. Slaked lime or milk of lime should be discharged directly from the slaker or lime feed tank into the water being treated. If, however, it is necessary to transport it any distance from the slaker or lime feed tank the following precautions should be observed:

1. The ejector water, used to maintain adequate velocity to prevent settling in the lime line, should be softened water. Hard water reacts with lime and produces a hard adherent scale in the lines and soon plugs them.
2. A trough with an easy-to-remove top is preferable to a pipe for conveying milk of lime. If a trough cannot be used, a rubber

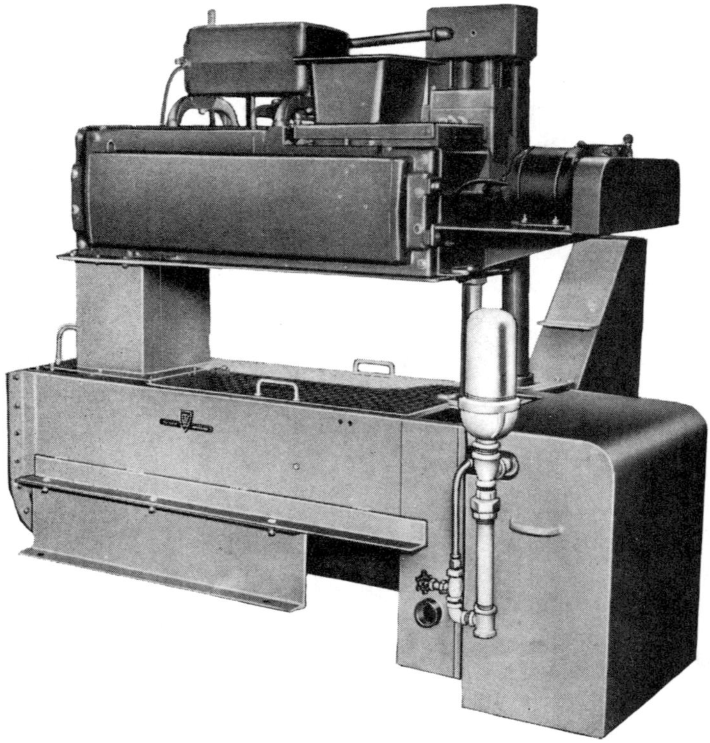

PICTURE 2-6. LIME SLAKER, PASTE TYPE, WITH LIME FEEDER ON TOP

hose, which is easily accessible for pressing or breaking down the scale in it, is preferable to a metal pipe.

Milk of lime is difficult to transport because the lime, being in suspension, naturally settles out. When this pasty calcium hydroxide cements itself with the siliceous residue from the stone, the combination produces a compound that is almost as hard as cement and exceedingly difficult to remove.

### *Direct Feed of Powdered Quicklime*

It is the usual practice to slake the quicklime in mechanical units, then apply the milk of lime slurry to the water. However, one plant* has used pulverized quicklime without slaking; reports indicate successful operation and economy over the use of hydrated lime. The finely pulverized CaO was fed with the usual hydrated lime feeder, and applied to the water by an eductor. Some sludge was recirculated through the feeder to minimize scale production in the slurry lines. The lime used in this plant was of such size that 100% would pass a 20 mesh screen,

*Water and Sewage Works, Vol. 102, p. 503, Dec., 1955.

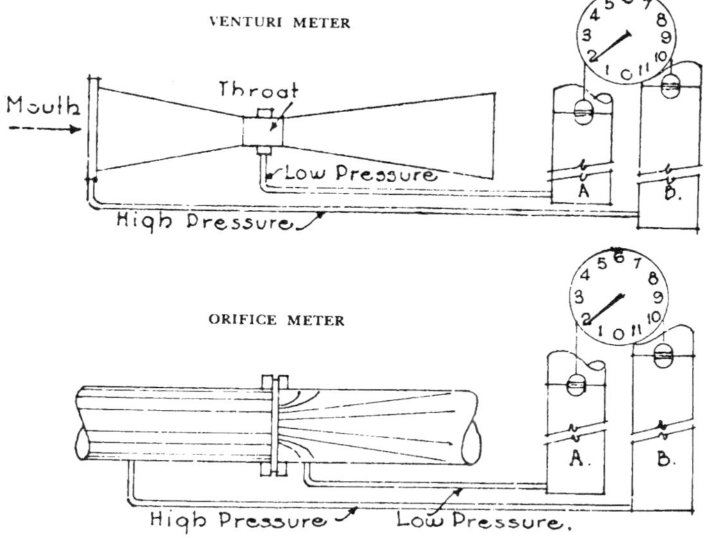


The pressure of the water in the pipe connected to the throat will be less than that in the pipe connected to the mouth of the tube. This differential in pressure changes proportionally with a change in the rate of flow of water through the tube. A device is provided for automatically interpreting the difference in pressure as indicated by the float tubes A and B so that the operator reads directly the rate of flow through the tube and the total amount of water that has passed the tube.

FIGURE 2-12. OPERATION OF METERS

and 90% would pass a 100 mesh screen. It is anticipated that further developments in method may be forthcoming.

## Measurement of Water

It is important that the amount of water being treated is known and recorded for maintaining uniformity of treatment. The selection and maintenance of accurate metering devices are important factors in water treatment plant operation. New principles of metering have been developed, which are accurate and easily maintained.

Common types of devices used for measuring large flows of water are the Venturi, orifice type, velocity, magnetic flow meter, ultrasonic type, propeller, rotameter, and flumes and weirs.

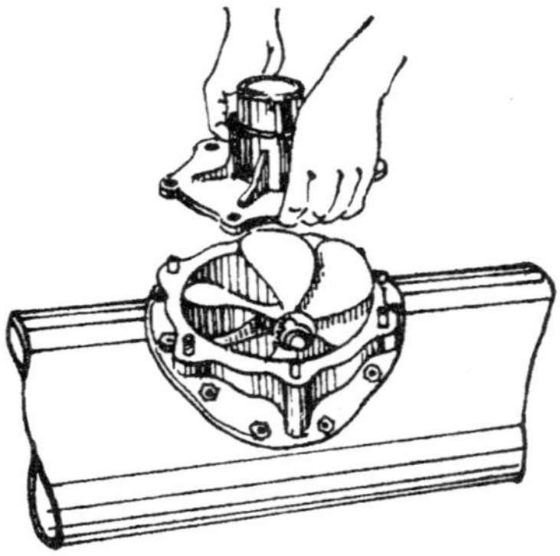

FIGURE 2-13. THE VELOCITY METER

The Venturi meter is a Venturi tube (a pipe with a streamlined constriction), float tubes, connecting piping, floats, and gauge, arranged generally as shown in Fig. 2-12. The orifice meter is similar in principle to the Venturi tube, the difference being the method of constructing the flow.

The magnetic flow type, one of the newer developments in water flow measurement, meters the flow by the changes in electrical conductivity, between two electrodes, as the velocity varies. See photo below.

The ultrasonic flow meter measures the time it takes for an ultrasonic pulse to travel across a pipe or conduit with and against the flow. Because it straps to the outside of the pipe, it is easy to retrofit and maintain. Finally, the propeller or turbine meter, the rotameter, and flowes and weirs are other devices use for measuring flow.

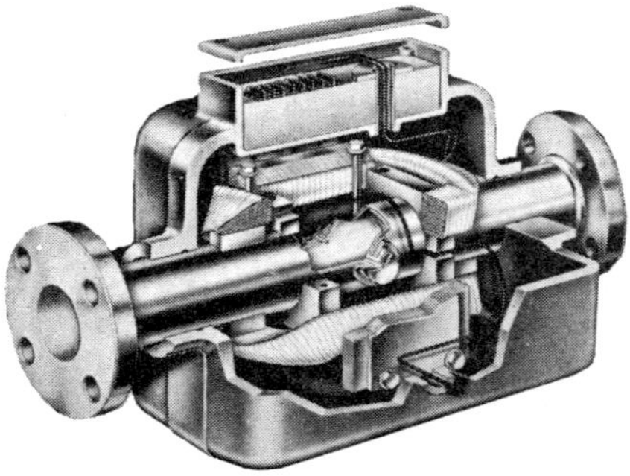

(COURTESY FISCHER & PORTER CO.)

MAGNETIC FLOW METER

The Velocity Meter (See Fig. 2-13) consists of a propeller, about eight- tenths of the pipe diameter, which rotates on a shaft. This shaft, geared to a vertical shaft, operates a register counter which records the flow of water through the pipe.

Chapter III

# COAGULATION AND SEDIMENTATION

## Coagulation

Almost all surface water supplies are turbid, colored, and contain bacteria. The turbidity is due to finely divided matter held in suspension. It usually is too light to settle unless allowed long periods of time. Color is caused by dissolved organic material resulting from algae, decayed vegetation and bacteria. Ordinarily these impurities may be removed readily by the use of coagulants. However, surface supplies which receive sewage and/or industrial wastes may require specialized treatment; first, because of the presence of detergents, which retard the coagulation process, and, secondly, because of color-bearing substances which may not be affected by coagulation.

The reactions involved in coagulation have been the subject of much research. Studies in recent years have broadened the original concepts to cover the principles of neutralization of the electrostatic charges on the suspended colloidal particles that are to be removed by the process. The dissociating coagulating chemicals, when added to water possessing natural alkalinity, produce a gelatinous, sticky precipitate. However, for the suspended colloidal particles to be effectively entangled or absorbed in the "floc" that the coagulant forms, it is necessary that their mutually repellent "like" charges be neutralized either by the action of the coagulant or coagulant aid.

The nature and extent of the electrostatic charges on the colloidal suspended particles in the water are determined by the measurement of their mobility or "zeta potential". This measurement is beyond the scope of many laboratories, but if it can be determined, it will assist in the selection of the types of coagulant and/or coagulant aid.

In introducing the coagulant to the water, a rapid mixing action of 15 to 45 seconds is needed to distribute the chemicals properly; then, a moderately slow agitation (15–60 minutes) may be used to allow the gelatinous mass to form the floc or clumps which will settle easily. Some coagulant aids are most effective when added a few minutes after the original dosage.

The coagulants generally used are aluminum sulfate, alum- polymer blends, iron-polymer blends, ferrous sulfate and lime, ferric chloride, ferric sulfate, and sodium aluminate and lime. Others have been used but not to any great extent. Coagulant "aids" are varied in nature, and include the following: magnesium chloride, sodium aluminate, activated silica, starch, and numerous high molecular weight polyelectrolytes. All the following can have an effect on the coagulation:

1. Amount of turbidity.
2. Type of suspended and/or colloidal particles
3. Coagulant or coagulants used
4. Coagulant aid used
5. pH value
6. Alkalinity
7. Size and uniformity of the particles
8. Dissolved solids, especially as to positive and negative ions present
9. Degree of salinity
10. Order of addition of chemicals
11. Time and type of mixing, especially in the second or flocculation stage.

In coagulating clear water, it is sometimes helpful to add artificial turbidity. Bentonite and other clays have been effectively used. When these clays are suspended in water, a nucleus is provided for floc formation. Experience on cold, low turbidity Lake Erie water indicates that a dosage of about 10 milligrams per liter bentonite is effective.

## Aluminum Sulfate

Aluminum sulfate was perhaps one of the earliest coagulants used in water treatment, and is now the most extensively used. It is made from bauxite and sulfuric acid.

$$\underset{\text{Bauxite}}{2Al(OH)_3} + \underset{\text{Sulfuric acid}}{3H_2SO_4} \rightarrow \underset{\text{Aluminum Sulfate}}{Al_2(SO_4)_3} + \underset{\text{Water}}{6H_2O}$$

The commercial product is slightly basic; that is, it contains some alumina in excess of that necessary to combine with the acid. The $Al_2SO_4$ content is about 17 per cent.

The general formula is $Al_2(SO_4)_3 \cdot 14H_2O$. High basic aluminum sulfate is probably $2Al_2(SO_4)_3 \cdot Al_2(SO_4)_2(OH)_2 \cdot 13H_2O$. It usually can be relied upon to give good results if the proper mix and conditioning are provided. The zone of best floc formation is at a lower pH than the iron salts. If used at its optimum pH of coagulation, low residual metal is found. It is very effective for reducing the normal carbonates, in many cases more than the theoretical amount. This property makes it especially valuable in water softening plants. This same property will, however, make most surface supplies take on an aggressive tendency - that is; the water will become corrosive.

## Alum-Polymer Blends

Within the past decade, technology has developed which allows certain cationic polymers to be blended with Alum in specific proportions at the factory, and shipped to the water plant for use as an homogeneous product. The floc formed by these products is similar to Alum floc, but can generally be counted on to be tougher and faster settling at dosages of only one-third that of Alum. This has produced longer filter runs, less sludge, and a more stable effluent with respect to calcium carbonate saturation. (See Journal AWWA, Confluence, April 1988, page 12).

## Iron-Polymer Blends

Similar advantages to the Alum blends are had with Iron blended polymers, and are used generally where iron salts are being used. The iron blends work well in waters where low alkalinities are encountered; that is, below 60 mg/L. Above these quantities of alkalinity, Alum blends work better, given the same mixing times and energies.

## Ferrous Sulfate and Lime

A combination of ferrous sulfate and lime is an effective coagulant for the clarification of turbid waters. Both materials are relatively low in cost. They may be used in various ratios. Ferrous sulfate may be used alone for secondary coagulation if there is sufficient oxygen in the water to insure complete oxidation Where prechlorination is practiced, chlorine may be used with ferrous sulfate instead of lime.

The reaction of chlorine with water furnishes the oxygen necessary to oxidize the iron. The reactions for this coagulant take place best at high pH values.

## Ferric Sulfate

Ferric sulfate as a commercial product for coagulation is comparatively new. Its wide pH zone of coagulation makes it applicable to various conditions. Iron salts do not form soluble compounds in the alkaline range, and for that reason very low residual metal is found at the high pH values. Except in special cases of color removal, coagulation with iron salts is best at a pH of 9.0 or above.

## Chlorinated Copperas

Ferrous sulfate is used as a base material for the preparation of chlorinated copperas. The ratio is one part of chlorine to eight parts of $FeSO_4 \cdot 7H_20$. In practice, usually about 10 per cent excess chlorine is used to insure complete oxidation. This material has proved effective for removing certain types of color in soft waters. Results are best accomplished at pH below 7, due perhaps to complex ferric organic precipitates that are formed. Lime is used to raise the pH to prevent corrosion, just before or after filtration. In the alkaline pH ranges, certain colors may be accentuated by the use of iron salts. Here also, complex ions, or colloids, of the iron and coloring matter are formed but are not precipitated.

## Ferric Chloride

Ferric chloride has a high optimum pH zone of coagulation but can be used in low pH ranges for color removal. Its greatest use has been as a precipitant and sludge conditioner in sewage treatment. Here a saving in ferric chloride on digested sludge is obtained by the use of lime in amounts sufficient to raise the pH to 9.0 or more.

## Sodium Aluminate

Sodium aluminate is more expensive than other coagulants. Its application for municipal plants has been limited. It is especially effective when used with lime and soda ash in the treatment of boiler feed waters. It forms a floc rapidly. The free alkali present gives it some softening properties. Unlike aluminum sulfate it will not convert normal carbonates into bicarbonates. Characteristics of the floc depend greatly upon the condition under which it is formed.

It has been found to be more effective under some conditions than others, and can result in savings in treatment costs if substituted for part of the alum dosage. It has been demonstrated to give a somewhat lower colloidal calcium carbonate content in the settled water than the usual alum coagulation will do when used in conjunction with lime and soda ash softening.

## Lime

Lime may be used as a coagulant for waters high in magnesium compounds. This is due to the magnesium hydroxide precipitated by excess lime. The floc of magnesium hydroxide has somewhat the same properties as aluminum and iron hydroxides. In softening plants where 20 to 30 mg/L of excess lime are carried, a saving can be effected by taking advantage of this coagulating material produced in the water.

Magnesium coagulant chemistry has been further developed by Dr. A. P. Black, a renowned water-chemist and his engineering associates. In the Black process two or three chemical compounds may be recovered for re cyclical use: the magnesium as a carbonate/hydroxide from the lime-treated raw water; lime from recalcination of the filtered calcium carbonate sludge; carbon dioxide gas from the recovery lime furnace for recarbonation. Use of alum or iron salts (coagulants), it is claimed, can be eliminated by using the recovered magnesium carbonate/hydroxide. Although this unique process potentiates significant economy, its commercialization has lagged, probably owing to it being an intricate process, with many steps, that is foreign to most water plants, particularly those that are small to medium sized. A high capital investment is necessary for the chemical plants that must be built. The reversible chemical reactions in obtaining the magnesium compounds are depicted below and the other applicable reactions involving causticization of magnesium compounds with lime are found in Chapter 10.

$$Mg(OH)_2 + 2CO_2 \leftrightarrow Mg(HCO_3)_2$$
$$Mg(OH)_2 + CO_2 \leftrightarrow MgCO_3 + H_2O$$
$$MgCO_3 + H_2O + CO_2 \leftrightarrow Mg(HCO_3)_2$$

## Polymers

Organic coagulant aids, often referred to as polymers, are very high molecular weight compounds of organic nature. There are many types available from an increasing number of suppliers; they may be purchased in solution or dry form and may be cationic, anionic, or nonionic in nature. The dry material often requires special precautions in wetting down and putting into soluble or slurry form.

Some suppliers claim that their products can replace all metallic salt coagulants, but for the most part it has been found that a small amount of the aluminum or iron salt is necessary for satisfactory and economical clarification. Using the polymer to replace all the alum or iron salts may require so much polymer that the cost is greater than the cost of metal salts originally used.

Personal experience indicates that the metallic salt dosages can be markedly reduced by the use of small amounts of some polymers, effecting an overall reduction in chemical costs. A reduction of lime and/or soda ash dosage also may be accomplished with some polymers. The polymers are quite expensive in cost per pound; but since the dosages are, as a rule, very low, the increase in costs for the polymer dosage may be more than offset by the decreased dosage costs for alum (or iron) salts as well as lime and/or soda ash.

It is strongly recommended that an evaluation of the various types and brands of polymers first be made in the laboratory using jar tests; then the most promising type or types be evaluated for both cost savings and effect on water quality on a plant scale. However, before using any polymers on a plant scale, the operator is warned that written evidence of approval by State and Federal regulatory agencies must be obtained, covering (a) the manufacturer, (b) the specific product used, and (c) the maximum quantities or dosages of that specific polymer that can be applied without involving any hazard to the water quality.

See Chapter 12 for a description of lime's function in precipitating trace amounts of heavy metals, some of which are toxic, along with radioactivity that are at unsafe levels in some surface waters.

## Laboratory Jar Tests

There is no rule or set of rules for determining the quantity of coagulant required, or which one, or combination of them, is best suited to obtain maximum benefits. A very muddy water in which the turbidity is due to coarse particles will require less coagulant that a less muddy one in which the turbidity is due to fine particles. The presence of organic compounds, or too much or too little acidity or alkalinity (pH value), sometimes prevents coagulation reactions, or causes them to proceed to completion very slowly. Color may prove troublesome by combining with the coagulant to form soluble compounds, making the treated water even more highly colored than the raw water.

The surest way to determine the quantity and type of coagulant required to obtain best results is by laboratory trial tests. For a dosage in grains per gallon added to one gallon samples, weigh 65 grams of chemical into 1,000 milliliters of distilled water. Each milliliter of this solution added to one gallon of the water to be treated equals a dosage of 1 grain.

Since "grains per gallon" has been supplanted by "milligrams per liter", the stock coagulant can be made up to contain 1.0 gram per 1000 ml. Hence, one milliliter of this strength solution added to 1 liter of raw water is equivalent to 1mg/L dosage.

The unit of "pounds per million gallons" is sometimes used. To set up jar tests in this unit, the stock solution should contain 1.2 grams of chemical per liter of solution so that each 0.1 ml. of this stock added to one liter of water is equivalent to 1 lb. per million gallons dosage. (See following table).

### *Stock Solution Strengths for Jar Tests*

*Test Portions—1 Gallon*

| *Stock Solution in grams/Liter* | *1.0 ml. stock to one gallon equals* |
|---|---|
| 65.0 grams | 1.0 gpg |
| 3.8 grams | 1.0 mg/L |
| 0.456 grams | 1.0 lb/MG |
| 4.56 grams | 10.0 lb/MG |

*Test Portions—1 Liter*

| *Stock Solution in grams/Liter* | *1.0 ml. stock to one Liter equals* |
|---|---|
| 17.1 grams | 1 gpg |
| 1.0 grams | 1 mg/L |
| 10.0 grams | 10 mg/L |
| 12.0 grams | 100 lbs/MG |
| 0.12 grams | 1 lb/MG |

Thus, for the jar test using a series of six 1-gallon samples, add 1, 2, 3, 4, 5 and 6 milliliters, respectively, of the 65 gm/L coagulant "stock" solution. If using one liter samples, use the 17.1 mg/L stock solution. Each jar should be dosed with increasing amounts of coagulant so as to produce a bracketed range. For instance, if you expect a needed dosage to be in the range of 10-12 mg/L, you might try a series of dosages at 8, 9, 10, 11, 12, and 13 mg/L. If the expected dosage is not known, try 4, 8, 12, 16, and 18 mg/L. Then you can attempt a second set of jars from observations of the first. This second set will help to "fine tune" the system. It is best to obtain the services of a partner to help you dose each sample with coagulant simultaneously, because of the short (30 sec.) rapid mix sequence. If it takes a span of 15 seconds to dose the first one liter sample to the last, the test will produce poor results. Mix the samples rapidly for 30 seconds, using stirring device shown in Fig. 3.1. Next, stir gently for 30 minutes and observe floc formation. Let settle for 30 minutes then filter portion through a paper filter. The minimum dose that produces a bright, sparkling effluent from the paper filter is the quantity that is probably needed. To be certain, run a filterability index on the supernatant.

## Filterability Index

Filterability index, F.I., is a unitless ratio of the time it takes to pass a given volume of treated water through a filter to the time it takes to pass an equal volume of distilled water through the same filter. An F.I. of 1.0 is ideal, while a 1.3 or greater indicates poor treatment technique. For example, the calculation of F.I. for a sample might be:

(for 50 milliliter samples)

Time through the filter for Distilled water = 6.1 seconds

Time through the filter for Treated water = 6.8 seconds

F.I. = 6.8/6.1 = 1.1

Many operators still use settled water turbidity results obtained from jar tests as a guide for determining dosage control; however, the F.I. test is replacing settled turbidity as a more reliable indicator of good pretreated water. This author has personally seen a settled water turbidity of 3.5 for both raw and settled water on the same day; but I would not consider filtering the raw water without treatment. These two waters would have a different F.I. respective to each other. It is not uncommon to work with raw water that has such a low turbidity that any coagulation will render it more turbid before treatment through filtration. This does not mean that the water quality has been impaired.

In making comparative tests of different coagulants a very large number usually must be made. Tests should be made not only on each sample, as is, but also on the samples with acidity or alkalinity (pH value) adjusted to various ranges. Laboratory results may be made to more accurately predict the plant requirements if the jar tests are run on samples maintained at nearly the same temperature as the raw water. A convenient way to do this is to procure a metal tank or "water bath", approximately the same height as the jars; short enough to fit under the stirring device and wide enough to permit easy handling of the jars. Tap water is introduced through a small jet set at the bottom in a horizontal position to cause agitation, and an adjustable over-flow provides a means to maintain the desired water level.

## Mixing, Agitation, or Flocculation

After the chemical has been added to the water the two should be quickly and thoroughly mixed. Prolonged agitation is also desirable and often necessary. In the treatment of turbid water the precipitate of the coagulating compound should be kept in suspension until it has had sufficient time to collect into clusters a maximum number of the fine particles, which give the water its turbidity. Gentle velocities of from 0.6 to 1.5 feet per second are adequate. If the water and chemicals are given a quick mix ahead of prolonged gentle agitation, the time is usually only momentary and the process extremely violent. The time of gentle agitation varies depending upon the temperature and the nature of the water treated. From 15 minutes to one hour is the usual practice.

A laboratory stirring device is one of the most needed pieces of equipment in a water works laboratory. It is of especial use in coagulation, color removal, water softening reactions, and taste elimination studies, and for determining minimum quantities of coagulants

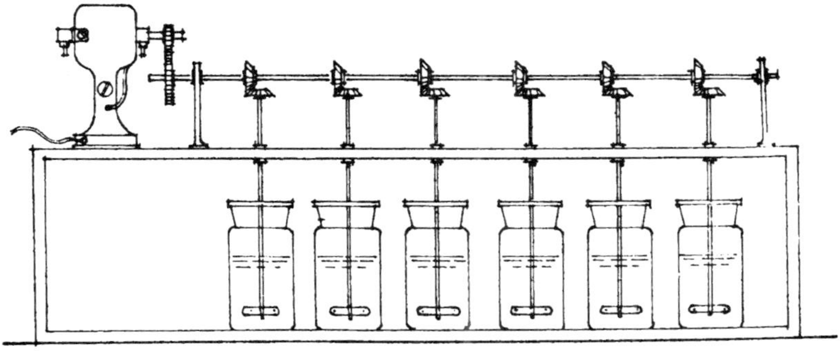

FIGURE 3-1. LABORATORY STIRRING DEVICE

FLOCCULATOR

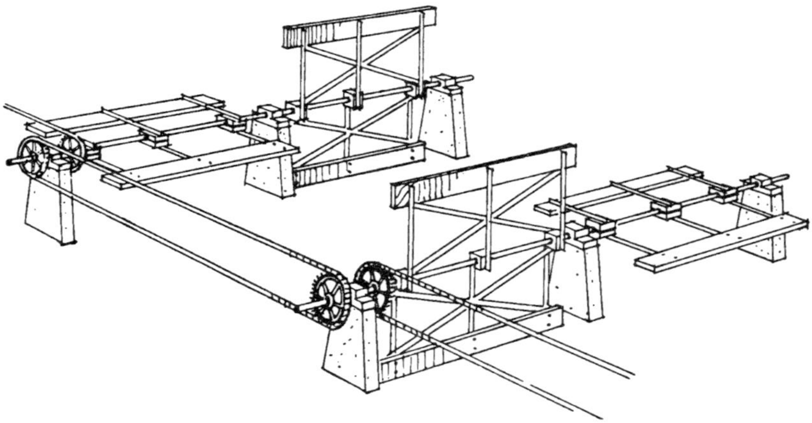

BAFFLED MIXING TANK

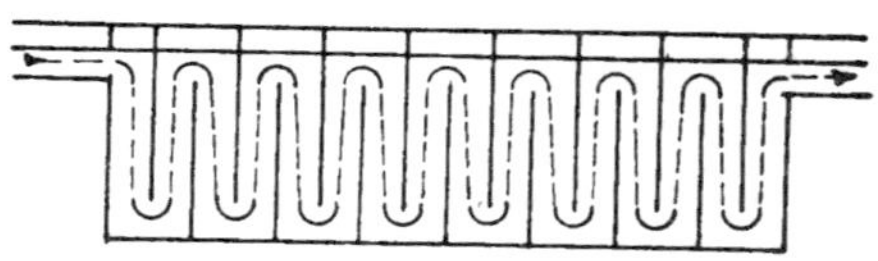

MECHANICAL AGITATOR

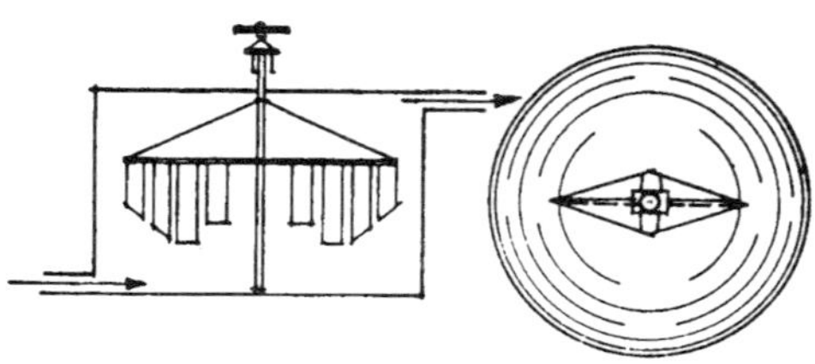

FIGURE 3-2. VIEWS OF MECHANICAL AGITATORS. APPEARANCE OF FLOC WHEN WATER ADMITTED IN BOTTOM SIDE OF TANK AND TAKEN OFF AT UPPER SIDE OF TANK.

required for treatment. The paddles usually rotate so as to produce a velocity of 1.5 feet per second.

The types of mixing devices used in water purification plants are baffled tanks, with various combinations; rapid mixing tanks provided with rotating paddles or propellers; diffused air supplied from a compressor; and the flocculator.

Most conventional water plants employ the "rapid mixing" and flocculation" steps as two distinct unit processes for proper coagulation. The main goal of rapid mixing is to introduce a chemical into the stream of raw water as quickly and evenly as possible. This chemical, or coagulant, usually carries an abundance of positive electrical charges which will neutralize, or destabilize the negative electrostatic charges which are associated with suspended turbidity particles. Without this destabilization, the suspended turbidity particles will not settle properly, nor will they be readily filtered. It should be noted that for a given water, there is an ideal rapid mixing time and chemical dose which, if not met, will have adverse effects on the process. These parameters can best be approximated through the aforementioned jar testing techniques.

Water which has been properly coagulated usually passes on to the flocculator where gentle stirring allows the tiny "floc" particles to grow in size, thus increasing their settleability. This form of coagulation is a physical process, as opposed to rapid mixing, which is considered to be a chemical process.

The flocculator consists of a horizontal shaft, the length of which is at right angles to the flow of water. Paddles attached to this shaft rotate so that an outward travel is given to the upper stratum of water being mixed, and an inward motion to the lower stratum. This inward motion causes the settled floc to be gently swept back along the bottom of the basin, lifted by the upward motion of the paddles, and mixed with the incoming treated water. The introduction of already precipitated floc is believed to facilitate coagulation.

Views of four types of agitating devices are shown in Figs. 3-2 and 3-3. Properly designed baffled tanks are satisfactory, but mechanical tanks are less expensive to build. They remove the necessity of lifting the water the extra two or three feet necessary to pass it through baffled mixing tanks. Although agitators require power to keep them running, the cost of operation, if proper drive is provided, should be less than the cost of the additional lift of the water. They also provide greater accessibility and flexibility.

Care should be exercised to insure uniform velocities throughout the tank. The paddle area should be at least 25 per cent of the area of the central vertical section of the tank, or 25 percent of the area of diameter of tank times its height. In the case of large tanks, when water is admitted through an inlet at the side or bottom of the tank and discharged at an outlet at the side and top, as is usually done, two distinct zones will develop, an inner and an outer one. The inner zone is from the center to about one third out and the outer zone from there on to the wall of the tank. The velocity is too low in the inner zone and scarcely any floc can be seen, whereas in the outer zone very satisfactory floc formation is obtained. More uniform velocities are obtained and short circuiting is minimized if the water is discharged through a take-off pipe from the center rather than the side of the tank.

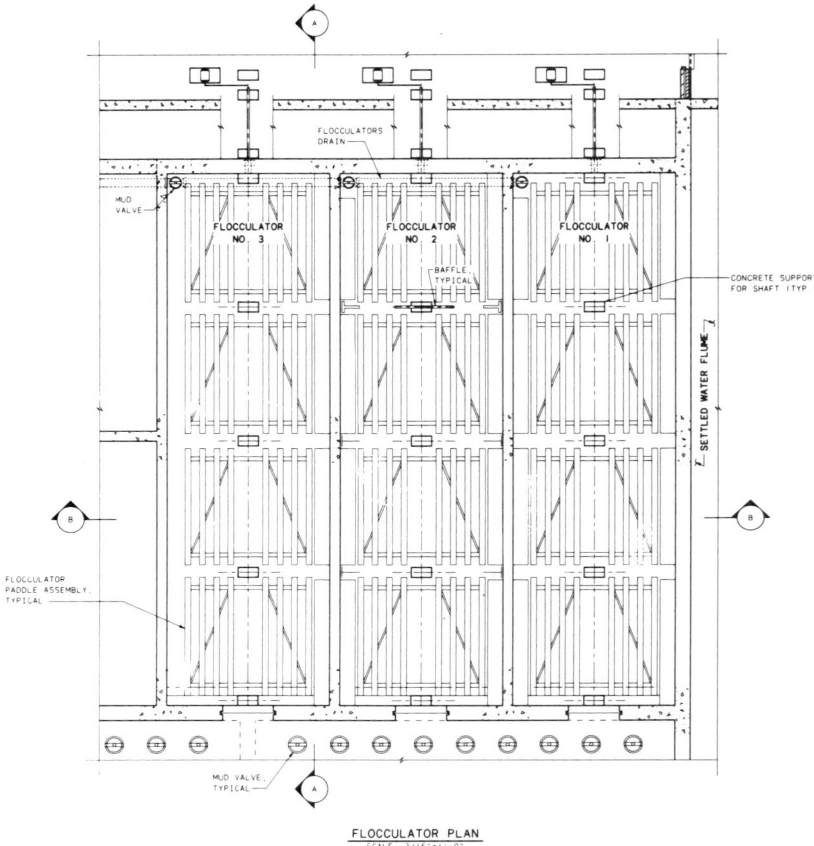


FIGURE 3-3A.

FIGURE 3-3A, B, C. MODERN FLOCCULATION SYSTEM SHOWING COMPARTMENTALIZATION. COURTESY BURGESS AND NIPLE, LTD. COLUMBUS, OHIO. THE FOUR BAFFLED COMPARTMENTS PRODUCE THE EFFECT OF INCREASING THE APPARENT DETENTION TIME OF THE TANK, THEREBY SAVING CONSTRUCTION SPACE. BY DECREASING THE NUMBER OF PADDLES FROM FOUR TO THREE TO TWO IN SUCCESSIVE COMPARTMENTS, A GRADUAL REDUCTION IN VELOCITY IS CREATED AS DESIRED WITH ONLY ONE DRIVE SHAFT.

## Sedimentation

Sedimentation and sedimentation basins are terms synonymous with settling and settling basins. Water works engineers and operators, however, think of a settling basin as one in which suspended matter settles without the aid of chemical coagulation, and a sedimentation basin as one in which settling is aided by coagulation. Sedimentation basins may

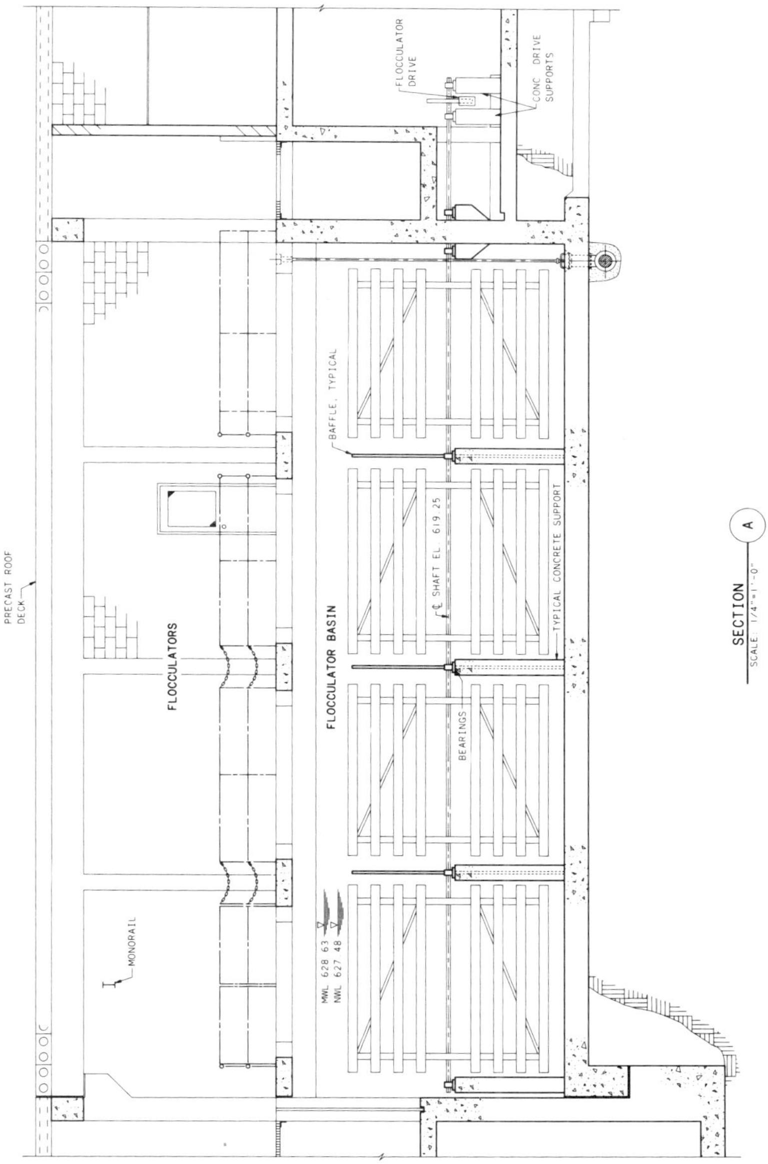
FLOCCULATOR DRIVE
CONC DRIVE SUPPORTS
BAFFLE, TYPICAL
℄ SHAFT EL 619.25
TYPICAL CONCRETE SUPPORT
PRECAST ROOF DECK
FLOCCULATORS
FLOCCULATOR BASIN
BEARINGS
MONORAIL
MWL 628.63
NWL 627.48
SECTION
SCALE 1/4"=1'-0"
A

FIGURE 3-3B.

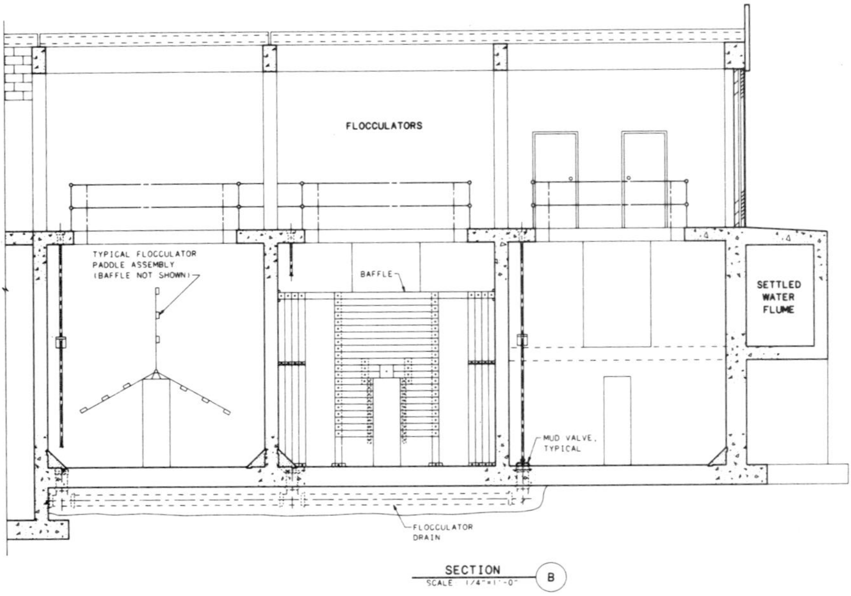


FIGURE 3-3C.

be operated by fill and draw, or by continuous flow. Almost all plants operate by continuous flow.

Modern sedimentation basins usually are about 17 feet deep and are square, round or rectangular in shape. Design is a much discussed subject among water works engineers. There is perhaps more disagreement regarding these basins, and less known about what actually happens in them than about any other part of a water plant. Sedimentation tanks for industrial and railroad water softening purposes generally are built vertically instead of horizontally, and sometimes are more than 50 feet in height. They usually are rated according to their detention period when operating at normal capacity. If a plant's capacity is one million gallons in twenty-four hours and the sedimentation basin holds two hundred and fifty thousand gallons, its detention period is six hours. In other words, when operating at normal capacity, six hours will be required for the water to pass through the basin. Six hours is about the average for the larger plants of this country. The increasing use of high molecular weight filter aids has shown that good filtration can be accomplished with very short detention times (see chapter on filters). The dimensions, length, width, and sometimes even the depth selected for basins, after the desired volume has been determined, are many times made on the basis of how they will fit into the general layout of the plant. Sedimentation according to some engineers and plant operators, depends on surface area and not depth. They advocate shallow basins. Others contend that sedimentation depends both on surface area and depth, the latter being of importance because it allows time and provides space for continued flocculation of suspended matter. Many older plants provided with inadequate mixing tanks, but deep coagulating basins (20 to 25 feet), produce satisfactorily settled water no doubt because of the depth of the basins. In other words, if water is not thoroughly coagulated before it reaches the basin, the depth is then a distinct advantage.

The design of sedimentation units has been the subject of much research, especially in regard to improving the quality of the effluent with reduced detention times. One company (Micro- Floc) has patented a unique "tube settler" device which can be installed in old or new clarifiers of any shape. These units are a radical change from the conventional design but are said to be very efficient. (See Figs. 3-4 and 3-5.)

Two basins usually are provided for small plants; three or more for larger ones. It is advisable to make provisions so that one can be out of service for cleaning or repairs. while the remainder can be operated. The velocity of flow through a basin usually does not exceed one foot per minute. It should be as uniform as possible. Figure 3.6 shows one inlet arrangement which has been successfully employed.

Neptune Tube Settler tube modules are formed by placing extruded plastic channels at 60° inclination between plastic sheets. By alternating the direction of inclination of each row, the module becomes a self-supporting beam which needs support only at the ends.

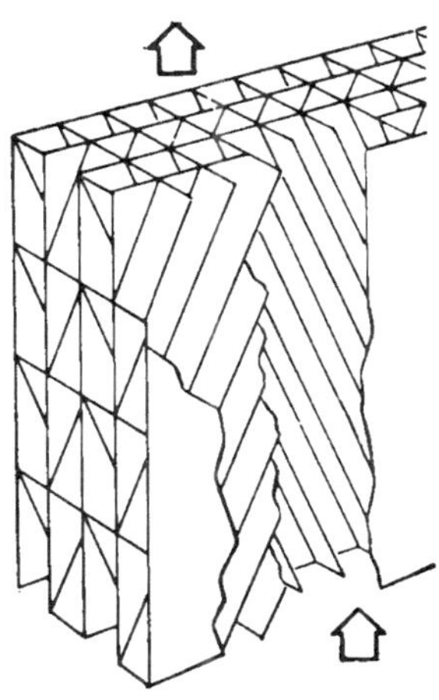

FIGURE 3-4. NEPTUNE TUBE SETTLER

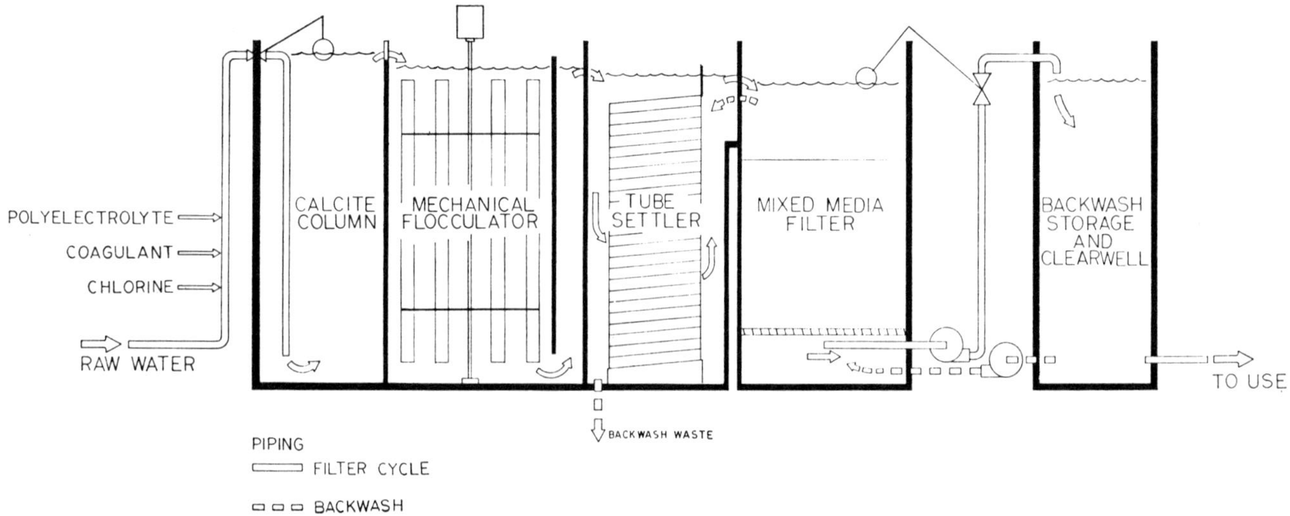


FIGURE. 3-5. FLOWSHEET INCORPORATING NEPTUNE TUBE SETTLER FUNDAMENTALLY THESE MACHINES ARE SIMILAR IN THAT THE PRECIPITATION OF SOLIDS OCCURS IN THE PRESENCE OF A LARGE AMOUNT OF PREVIOUSLY FORMED SLUDGE, AND THEY COMBINE FLOCCULATION OR MIXING, SETTLING, AND CONTINUOUS REMOVAL OF SLUDGE IN ONE TANK.

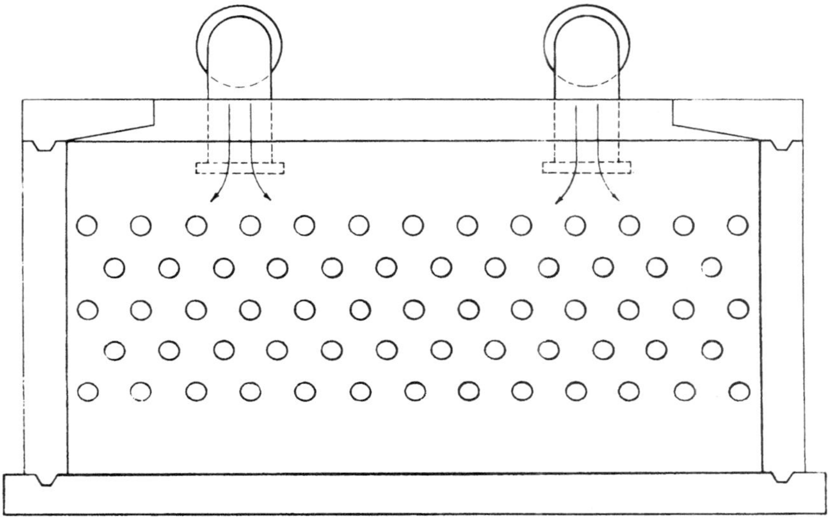

FIGURE 3-6. BASIN INLET ARRANGEMENT AT LAKE COUNTY, OHIO WATER TREATMENT EAST PLANT.

Water may enter the sedimentation basin through pipes or channels traveling at a much greater velocity than it will or should have going through the basin. This velocity must be reduced at the inlet of the basin and the water started through the basin at a slow uniform velocity over the entire path of its flow. In circular tanks this is accomplished by admitting the water into a circular central well extending from slightly above the surface of the water to within a few feet of the bottom. The water discharges from the bottom and is collected over a weir extending around the entire circumference of the tank. In rectangular and square tanks, perforated or slatted baffles may be built across the inlet and outlet ends of the tank. These baffles generally are located about three feet out from the basin walls.

Combined coagulation and settling should produce a water with a turbidity of under 5 NTU; in some plants turbidities as low as 0.5 NTU are achieved in the settled water. In the treatment of flood water this often means a reduction in turbidity of over 99 per cent. The sludge produced by coagulation or softening may be allowed to accumulate in the basin. It can be discharged intermittently or removed continuously by special equipment which has been developed for that purpose. Almost all softening plants are equipped with continuous sludge-removing devices because the quantity of sludge produced by softening is

PICTURE 3-1. FLOATING SIPHON SLUDGE COLLECTORS AT LAKE COUNTY, OHIO WATER TREATMENT EAST PLANT.

often twenty times as much as that produced by coagulation. There has been a trend to equip newly built surface water plants with such equipment.

Tanks designed for intermittent cleaning are provided with bottoms having steep slopes to an outlet gutter, or to outlet sumps distributed over the area of the floor. The sludge is allowed to accumulate for a period varying from several weeks to several months before a basin is cut out of service for cleaning. There are two general types of mechanisms for the continuous removal of sludge. They are illustrated in Fig. 3-7 and Pic. 3-1.

The disadvantages of allowing the sludge to accumulate are:

1. Each successive day the capacity of the basin decreases, thus cutting down its detention period.
2. Periodic shutdowns are necessary and cleaning is a messy, laborious job.
3. Its presence in a basin over long periods of time may cause tastes and odors to develop.

Some sedimentation basins are open, others have covers. Covers are generally used in cold climates where difficulties are apt to be encountered due to freezing, or where smoke, dust, and other objectionable materials are likely to contaminate the water.

## Sludge Blanket Filtration Devices

There are several different units now being actively marketed, which employ upward flow through a sludge blanket for water softening or clarification. These are:

1. The Spaulding Precipitator; The Permutit Co., New York, NY.
2. The Accelerator, Infilco-Degremont, Inc., Richmond, Va.
3. The Hydro-Treator; Dorr-Oliver, Inc., Stamford, Conn.

Coagulation and water softening reactions are hastened due to the contact of the chemically treated water with tremendous areas of sludge, and clarification is promoted because of the upward filtration

RAKE TYPE CONTINUOUS SLUDGE-REMOVING DEVICE

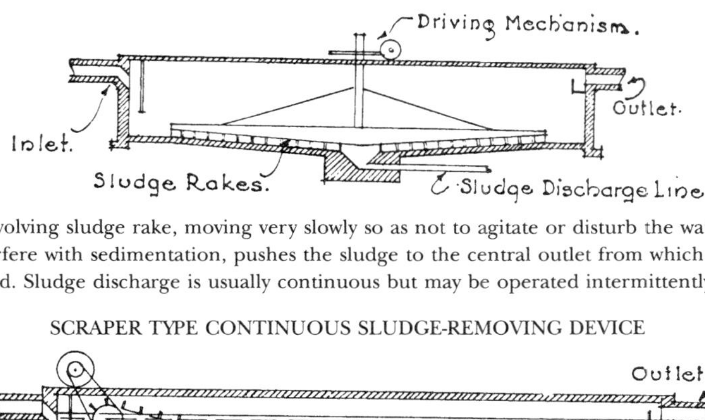


The revolving sludge rake, moving very slowly so as not to agitate or disturb the water and thus interfere with sedimentation, pushes the sludge to the central outlet from which it is discharged. Sludge discharge is usually continuous but may be operated intermittently.

SCRAPER TYPE CONTINUOUS SLUDGE-REMOVING DEVICE

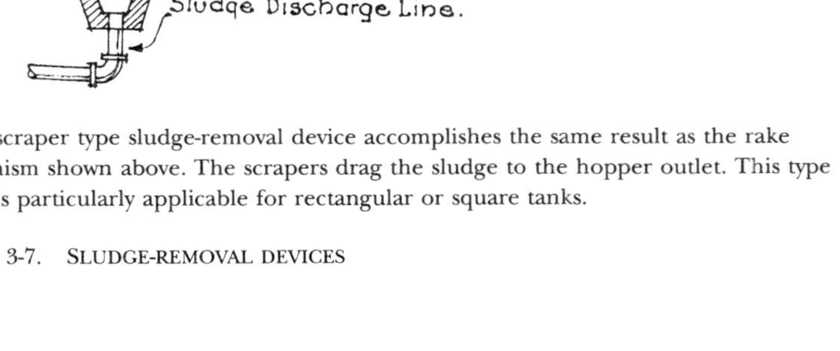


The scraper type sludge-removal device accomplishes the same result as the rake mechanism shown above. The scrapers drag the sludge to the hopper outlet. This type device is particularly applicable for rectangular or square tanks.

FIGURE 3-7. SLUDGE-REMOVAL DEVICES

of the coagulated water through the sludge blanket. As a result, the surface area of the basin can be reduced. The design rate varies from about 1 gallon per square foot per minute to about 2 gallons per square foot per minute, with mixing and flocculation lasting 15 to 30 minutes depending on temperature.

In some instances the cost of these devices, patent fees included, has exceeded the cost of the larger conventional type basins. In other instances they have cost less.

Control of the sludge blanket level may be difficult, particularly where fluctuations in raw water quality require frequent changes in chemical dosages. Also, best performance is obtained when these units are operated at a uniform rate. After the operator learns by actual experience just what he can do with a unit of this type, excellent results are to be expected. The marketed devices are shown in the accompanying sketches in Figs. 3-8 to 3-10.

## The Spiractor

The Spiractor is a recent development in lime-soda ash water softening equipment. It consists of an inverted conical tank in which the softening reactions take place in the presence of a suspended bed of granular calcium carbonate.

The advantages claimed for this type of softener are:

1. Equipment is small, 5 to 10 minutes detention period only is required.

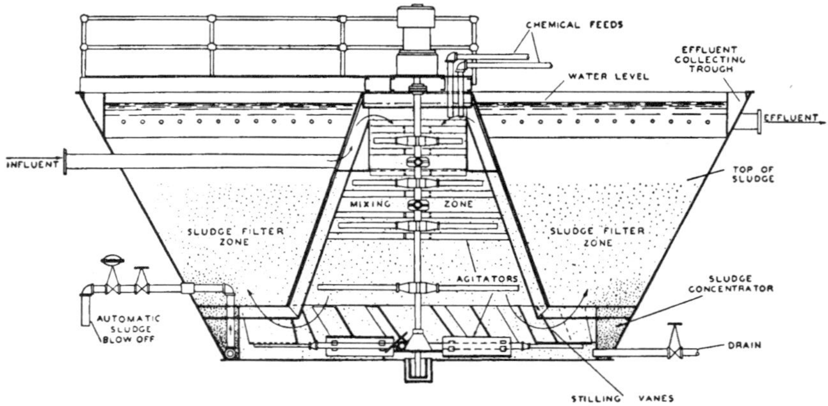


FIGURE 3-8. COLD LIME-SODA ASH PRECIPITATOR WATER SOFTENER

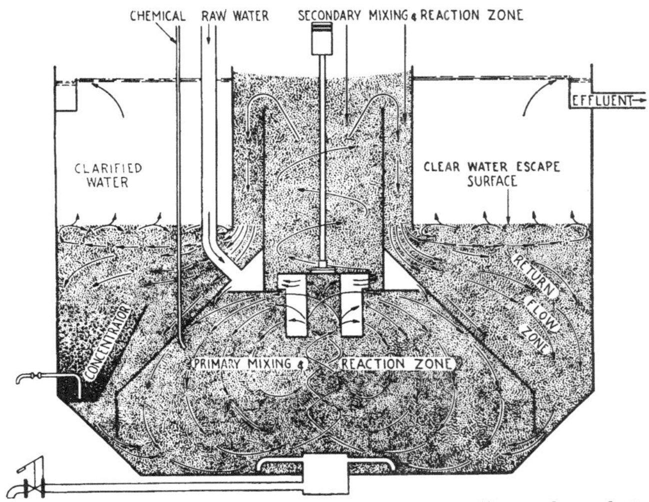


(COURTESY INFILCO, INC.)

FIGURE 3-9. COLD LIME-SODA ASH ACCELERATOR WATER SOFTENER

2. The precipitant and removed hardness are discharged in a granular form instead of the usual wet, pasty sludge. The granular material is easy to handle as it drains to less than 5 per cent moisture in a relatively short time.

The process involves the principle of catalytic precipitation, the granules serving as nuclei around which the precipitated compounds of calcium and magnesium build up.

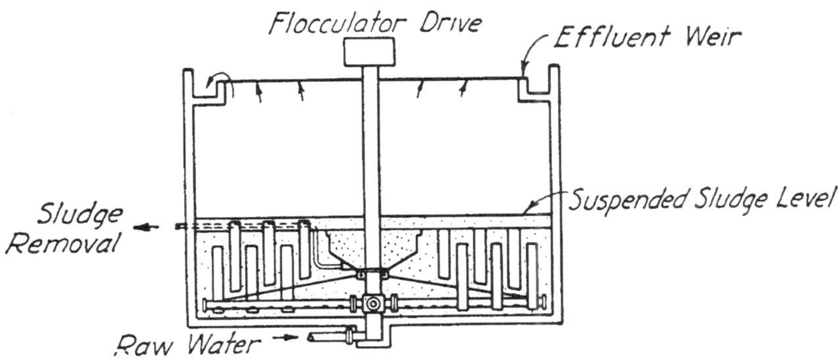


FIGURE 3-10. COLD LIME-SODA ASH HYDRO-TREATOR WATER SOFTENER

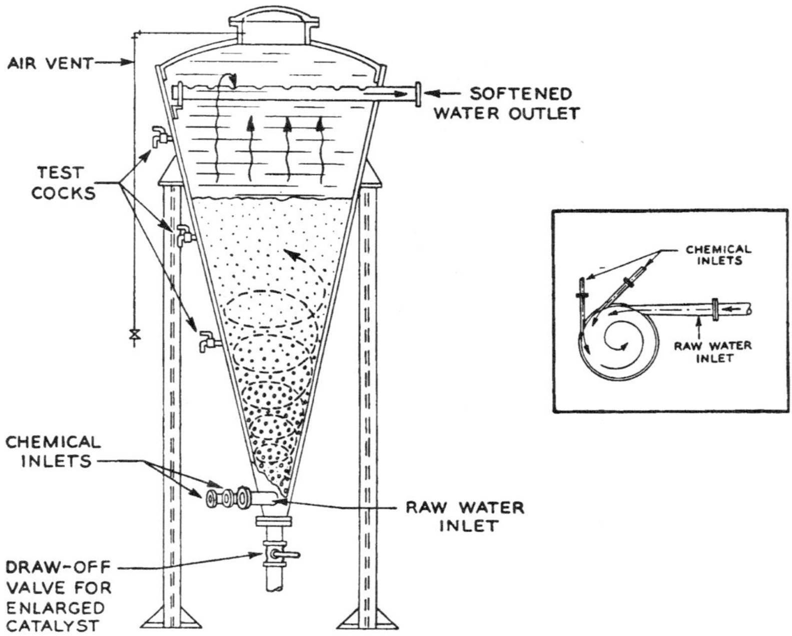


FIGURE 3-11. CROSS-SECTIONAL VIEW OF PERMUTIT SPIRACTOR SHOWING FLOW. INSET GIVES DETAIL OF ARRANGEMENT OF RAW WATER AND CHEMICAL INLETS. THE WATER DISCHARGED FROM THE SPIRACTOR USUALLY IS STABILIZED AND FILTERED THROUGH A SAND FILTER.

In operation, the conical-shaped tank (see Fig. 3-11) is slightly more than half filled with fine granules of calcium carbonate or sand (0.1 to 0.2 mm. in size). Raw water enters the bottom through a special fitting containing a tangentially disposed nozzle. Through another opening in this fitting the dosage of lime or lime-soda ash suspension is introduced. The chemicals mix with the upward swirling water and in the presence of the calcium carbonate granules the softening reactions take place—a contact reaction similar to that which occurs in sand filters softening unstabilized lime-soda softened water. The calcium carbonate granules grow in size just as filter sand does. As the granules grow in size and the bulk of material increases, its level is kept down to that desired by withdrawals of the largest particles from the bottom. Periodic additions of new material are required, and this new material may be produced by regrinding and screening a portion of the discharged material.

Chapter IV

# Filters and Filtration

## The Mechanism of Filtration

Filtration can be defined as the removal of suspended matter, (either naturally occurring or created from the treatment process), by passage of the water through a granular or porous medium. The efficiency of this removal is dependent upon two mechanisms: transport and attachment. The grains of anthracite or sand in a filter are considered to be collectors; that is, they gather floc particles from the influent suspension. These floc particles must be transported from the water to the collectors by gravity, diffusion, or some other hydrodynamic property. As the water flows quickly by the collectors, streamlines of flow are created which must be overcome by the floc particles. If these particles are successful in overcoming the streamline forces, then some sort of attachment process must occur as the particle nears the collector. The particle will "stick" to the collector only if the natural repellent forces associated with the floc particle have been stripped away, or destabilized, in the process of coagulation and flocculation. The best constructed filters will not collect floc particles from poorly coagulated water.

As floc particles attach to the filter media, they in turn become collectors. This is known as the process of filter ripening, and can explain why a newly cleaned filter will pass some turbidity particles initially. As the filter ripens, it becomes more efficient in producing a clear effluent. Until then it may be wise to filter to waste if possible. The filter ripening process works in favor of the plant operator for the bulk of the run, but eventually, the collectors become so overburdened with floc that the interstitial spaces of the media close down and offer increasing resistance to flow (loss of head). The shear forces created then become so great that floc may be stripped from the collectors and swept into the effluent. The filter run must be terminated and the bed must be backwashed.

## Gravity or Rapid Type Filters

Filters of the gravity or open type commonly are built of concrete. When more than one are required, they usually are placed side by side in batteries (Fig. 3-3). A modern filter can be described as a concrete box or basin provided with a strainer system, covered with 18 to 24 inches of graded gravel, which supports from 24 to 30 inches of filter sand or mixed media. See Fig. 4-1. Coagulated, settled water filters down through the media and gravel. When the filter becomes clogged, it is washed by forcing filtered water upward through the strainer system, gravel, and sand. The media is lifted and thoroughly agitated in the rising stream of clean water. All the finely divided and suspended matter collected on the media is carried away in the wash water and wasted. After washing, the filter is again ready for service. The washing can last from eight to twelve minutes or more.

## How A Filter Operates (See Fig. 4-2)

1. Open valve A. (This allows water from settling basin to flow to filter.)
2. Open valve B. (This allows water to flow through filter to filtered water storage. During filter operation all other valves are closed.)

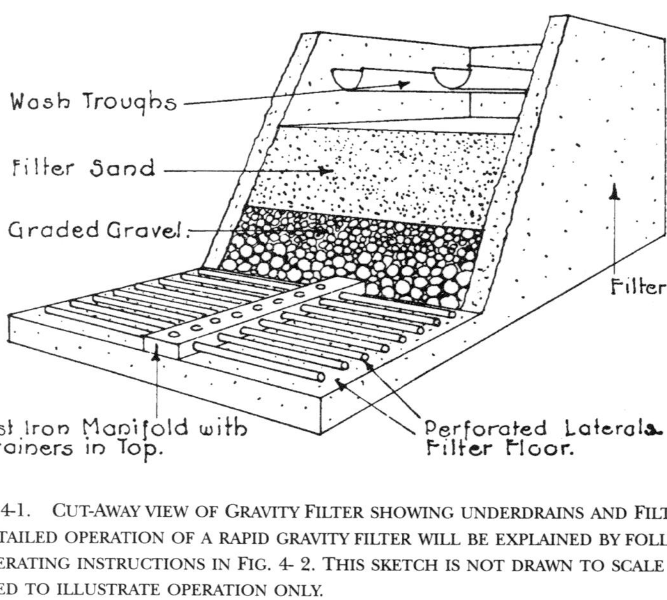


FIGURE 4-1. CUT-AWAY VIEW OF GRAVITY FILTER SHOWING UNDERDRAINS AND FILTER-BED. THE DETAILED OPERATION OF A RAPID GRAVITY FILTER WILL BE EXPLAINED BY FOLLOWING THE OPERATING INSTRUCTIONS IN FIG. 4- 2. THIS SKETCH IS NOT DRAWN TO SCALE AND IS INTENDED TO ILLUSTRATE OPERATION ONLY.

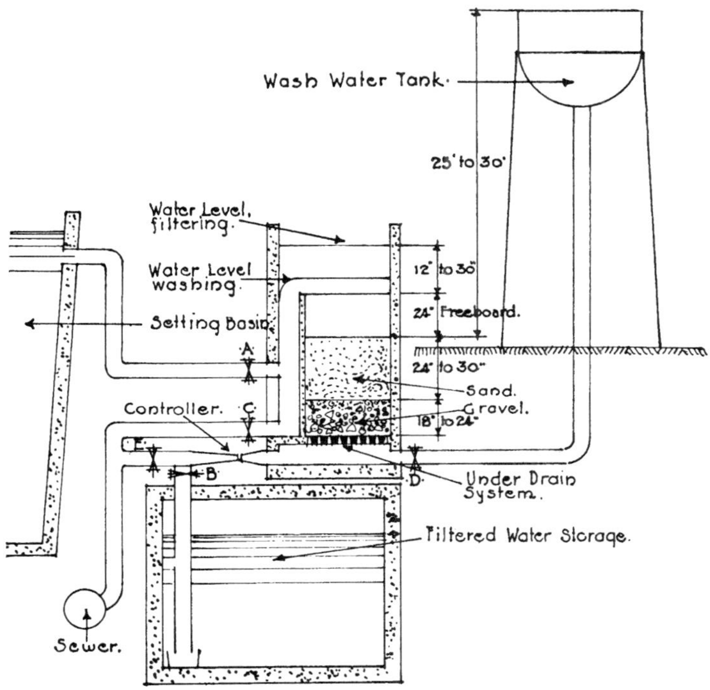


FIGURE 4-2. DIAGRAMMATIC SKETCH SHOWING FILTER OPERATION.

## How Filter Is Washed

1. Close valve A.
2. Close valve B when water in filter drops below top of overflow.
3. Open valves C and D. (This allows water from wash water tank to flow up through the gravel and sand, loosening up the sand and washing the accumulated dirt from the surface of the sand, out of the filter, and into the sewer.)

## How to Filter to Waste

1. Open valves A and E. All other valves closed. Water is sometimes filtered to waste for a few minutes after filter has been washed in order to condition the filter before it is put into service.

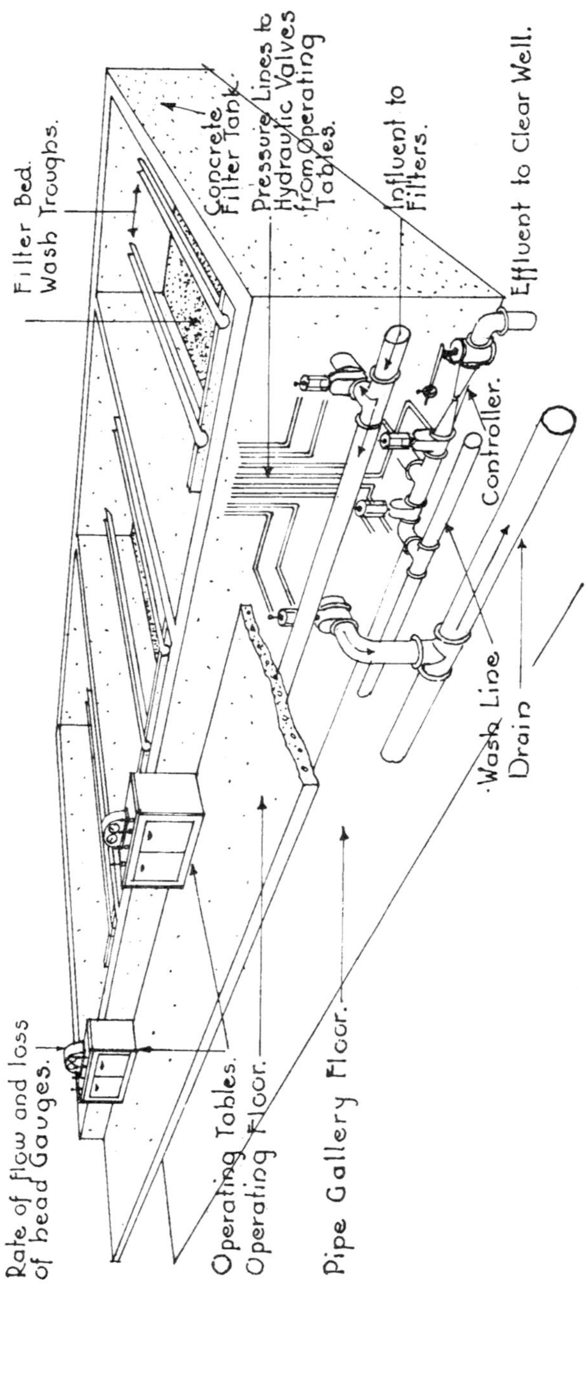


FIGURE 4-3. BATTERY OF THREE FILTERS

PICTURE 4-1. CENTRALIZED CONTROL PANEL FOR ENTIRE COCOA, FLA. WATER PLANT

A battery of three filters showing the general arrangement of filter, piping scheme, and operating control tables is shown in Fig. 4-3. A view made from a photograph of the interior of the filter room of the Cocoa, Fla., Water Softening and Purification Plant is shown above.

The most important appurtenances and parts to a filter are the device for measuring the loss of head, the device for controlling the rate of flow through it, the piping, the underdrain system, the gravel, the sand, and the wash-water troughs.

## Loss of Head

The loss of head measurement is important because it indicates whether or not the filter is operating satisfactorily and normally. The loss of head should rise steadily throughout the filter run. Any sudden drop in loss of head shows a breaking or cracking of the sand bed which might permit unsatisfactorily filtered water to pass through. It is used also to indicate when the filter needs to be washed. An explanation of loss of head operation is shown in Fig. 4-4. In the more modern filter plants loss of head is not measured by tubes and floats, as shown in this illustration. They have been replaced by the more modern differential pressure cells, but the scheme outlined shows the principles involved in a more simplified way and for that reason is used here.

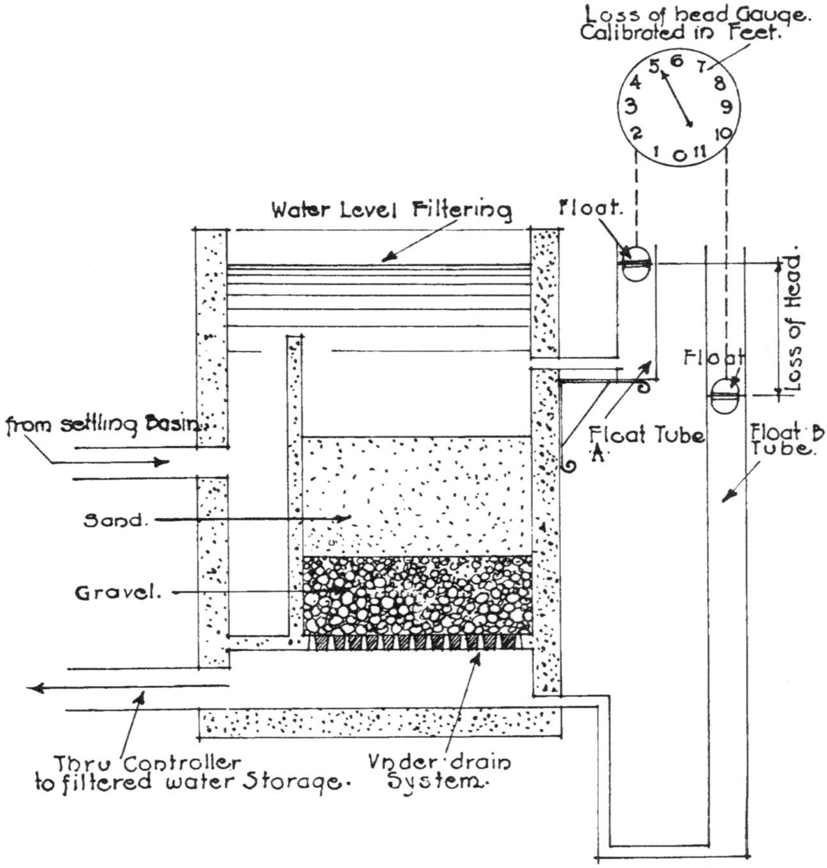


The water level in float tube A is always the same as the water level in the filter. When the filter is clean, the water level, for all practical purposes, is the same in B as in A (any slight difference is compensated for in gauge). As the sand becomes clogged it offers more resistance to water going through it and consequently the pressure in the collecting system becomes less, thus causing the water in float tube B to drop. The gauge above records the difference in feet. When the water in tube B drops to a point below the elevation of the underdrain and the filter is continued in service, it is said to be operating under negative loss of head. Water will continue to flow through the filter without having a positive head, or under a negative head, because of suction through the discharge pipe.

FIGURE 4-4. SKETCH SHOWING HOW LOSS OF HEAD OF GRAVITY FILTER IS MEASURED

Terminal head loss through most filters is usually 4 to 10 feet. The rate of loss for a multi media unit is much slower however than for a conventional rapid sand unit. This characteristic produces the larger filter runs associated with the mixed media units.

Care should be taken to ensure that the head loss at any level in the filter does not exceed the static head as this will produce a vacuum. Operators refer to this situation as negative head. This condition can result in air binding of the filter because, as pressure drops in the bed below atmospheric pressure, dissolved gases in the water escape and entrain in the media. The only remedy for a filter that is air bound is to give it a thorough wash. "Bumping", temporarily subjecting the filter to wash water, is unacceptable, and will only result in turbidity breakthrough.

## Rate of Flow Control

The device for controlling rate of flow of water through a filter is called the rate of flow controller. In normal operation this rate is two gallons per square foot per minute. It seldom is much less than the 2 gallon rate, but newer development in mixed media filter beds and use of filter aids allows the rate to be as high as 10 gallons per square foot per minute, with five or six gallon rates being frequently used.

If it were not for the rate of flow controller, the filtration rate would not be uniform. Water would go through the filter at a very high rate at the beginning of the run, perhaps not being well filtered, which would tend quickly to pack and clog the filtering material. On the other hand, the flow might be very low near the end of the run. The controller automatically allows the water to pass through the filter at a constant rate. Fig. 4-5 illustrates the principle of operation, although the actual construction of the rate of controller may be much different than the diagram shown.

The controller is calibrated so that for a given setting of the counterweight on the lever arm, the desired rate of flow will pass the balanced valve. When a filter is clean, water will tend to flow through it more rapidly than desired. As the flow through the Venturi tube increases, the difference in pressure between the upper and lower sides of the diaphragm increases. The greater pressure being on the upper side, the valve is lowered, throttling the flow until the predetermined rate, for which the machine is set, is reached. As the filter becomes clogged, the flow tends to decrease. This increases relatively, the pressure from the throat of the Venturi tube to the lower side of the

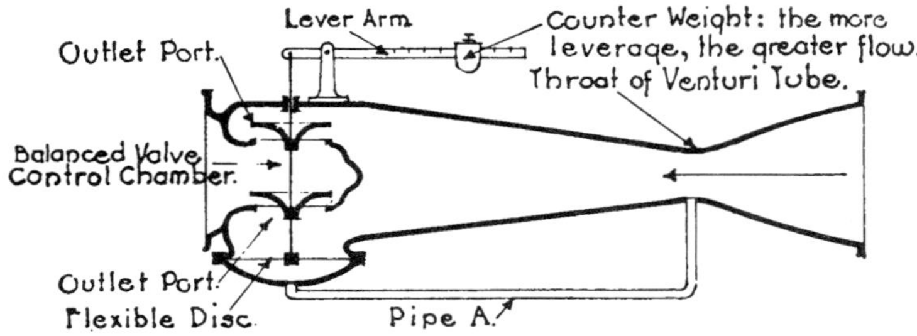


FIGURE 4-5. RATE OF FLOW CONTROLLER

diaphragm, causing the valve to rise (open) until equilibrium is again established at the predetermined rate for which the controller is set.

The rate of flow through a filter may be checked as follows:

1. Close the effluent valve.
2. Fill the filter to a marked elevation.
3. Close influent valve.
4. Open effluent valve.
5. Determine the time in minutes necessary to filter the water in the filter down one foot.

The filtration rate in GPM/ft² then is calculated as:

$$\frac{7.48 \text{ gal/ft}^3}{(\text{min/ft}) \text{ drained}} \times \frac{\text{water area ft}^2}{\text{sand area ft}^2}$$

## Filter Underdrain or Strainer System

The rapid sand filters in earlier years were built with an underdrain or strainer system consisting of a rectangular cast iron manifold or header in flanged sections, with bells along the side for insertion of laterals on six inch centers. (See cut-a-way view of Gravity Filter, Fig. 4-1). The laterals were closed at the ends but were drilled on the underside with orifices. These orifices were usually on six inch centers and of such a size that their combined area would be equivalent to about 0.33 per cent of the area of the filter surface.

Various types of patented underdrain systems have been developed

PICTURE 4-2. FILTER GALLERY. WATER SOFTENING AND PURIFICATION PLANT, CEDAR RAPIDS, IOWA

and have been in use for a number of years. The purpose of the underdrain system is to provide an outlet for the water passing through the filter and to admit wash water to the under side of the filter. It is extremely necessary that the rate of filtration be uniform over the entire area of the filter, and it is equally important that the application of wash water be uniform. Any of the various types available should be satisfactory provided they are covered with sufficient gravel to distribute uniformly the wash water, and graded so there will be no disturbance of the gravel layers.

## Filter Gravel

The filter bed usually consists of six layers of gravel, but one or more of the coarser layers are eliminated in case one of the patented false bottoms is used. The coarsest grade is on the bottom and the finest is on the top, upon which is superimposed a layer of graded and washed filter sand. The following grading and sizes of gravel have been found to be satisfactory:

| *Layer No.* | *Passing through Screen Openings in inches* | *Retained on Screen Openings in inches* | *Depth of Layer in inches* |
|---|---|---|---|
| 1 | $2\frac{1}{2}$ | $1\frac{1}{2}$ | 4 to 9 |
| 2 | $1\frac{1}{2}$ | $\frac{3}{4}$ | 5 |
| 3 | $\frac{3}{4}$ | $\frac{3}{8}$ | 4 |
| 4 | $\frac{3}{8}$ | $\frac{3}{16}$ | 3 |
| 5 | $\frac{3}{16}$ | $\frac{3}{32}$ | 2 |
| 6 | Torpedo sand 1.2 mm to 0.8 mm | | |

* Depth depends on type of strainer system.

The gravel should consist of hard, rounded, durable particles weighing approximately one hundred pounds to the cubic foot. It should be free from flat, thin, or long shaped pieces and washed and screened free from loam, sand, clay, shells, dirt, and other foreign material.

For best filter performance, the gravel size should be kept close to tolerance. Personal experience has indicated the need to sample and test each shipment. The quantity passing through the larger of the two screens, but retained on the smaller screen, should be at least 90%. For example, for the gravel used in layer No. 5 the amount retained on the 3/16 in. screen plus that passing the 3/32 in. screen should not exceed 10% of the sample tested, with an 8% value being much more suitable.

## Filter Sand

For monomedia filters, the layer of graded and washed filter sand should be 24 to 30 inches in depth. The sand usually has an effective size ranging from 0.35 to 0.50 millimeters and a uniformity coefficient of not more than 1.70. Effective size is defined as the size of grain in a sample such that 90 per cent = theoretical size of sieve (in millimeters) that will pass 10 per cent of sand. See Chapter XIX for procedures in regard to determining the effective size and uniformity coefficient. These values change for the multimedia filter and will be discussed later.

## Anthrafilt

Anthrafilt, a type of hard coal properly crushed to the desired effective size and uniformity coefficient, is being used successfully as a filter

PICTURE 4-3. FILTER GALLERY, WATER SOFTENING AND PURIFICATION PLANT, THE MAHONING VALLEY SANITARY DISTRICT, YOUNGSTOWN, OHIO

medium instead of sand in a number of filters. It weighs about 50 pounds per cubic foot (about half that of sand). Being lighter than sand, it has the advantage of requiring lower wash water velocities to expand it. This results in a considerable saving in wash water.

## Multimedia Filters

The first development in multimedia filters was a combination of sand and anthrafilt in which a top dressing of about eight inches of coarse anthrafilt on top of the sand acted as a roughing filter and increased the length of the runs. Research started during the mid 1940's led ultimately to more refinements in which several layers of filter media are used, starting with the finest sand next to the torpedo sand and ending with the coarsest material on top. The layers are kept in this position with little intermixing after backwash by careful selection of the proper density of materials for each layer. This type of filter, particularly if used in conjunction with the automatic controls and filter aid, as recommended, has resulted in a remarkable increase in both filter rates and length of runs with no decrease in quality of filter effluent. Incorporation of this type filter has enabled over-loaded plants to double their output at a savings of over half the cost of normal plant expansion. This system works because it greatly reduces head loss buildup over time.

## Mixed Media Filter

A top dressing of eight or more inches of anthrafilt over the sand was the original "mixed media" filter. However, the present mixed media may consist of a more elaborate system. One system (Micro-Floc), which has been meeting with remarkable success, uses a top layer of coarse anthrafilt, then a layer of low density graphitic sand of intermediate size, then a bottom layer of fine, high density garnet sand. The torpedo sand is also the dense garnet material. This coarse to fine media gradient is achieved by carefully choosing the specific gravities of the materials, as well as the effective sizes and uniformity coefficients. If properly selected, the materials will undergo substantial mixing between adjacent layers, giving rise to the term "mixed media". The intermixing provides for a more gradual change from coarse to fine particle, which is presumed to be ideal.

The coarser media at the top of this filter will allow for longer filter runs, while the smaller grain sized material at the bottom of the bed will produce a clearer filtrate. The top three layers, because of the density differences, tend in general to stratify according to density and, inversely, as to particle size after backwashing.

A filter of this type is said to be useful the entire depth, selectively removing the coarse, intermediate, then the fine suspended matter as the flow progresses downward. When used in conjunction with suitable polyelectrolyte filter aids and automated control of both coagulants and filter aids by continuous monitoring of turbidities, these systems can permit much greater filter rates, even with reduced sedimentation times. A number of these filters are running at 5 to 8 gallons per square foot per minute.

## Pilot Filtration

A pilot filtration study of multi-media filters has been conducted at the Lake County, Ohio, Aquarius Water Plant.

Picture 4-4 is a depiction of the pilot filtration apparatus employed at the Aquarius Filtration Plant. An accompanying schematic of the assembly is shown in Fig. 4-6. These units are 5 inch diameter plexiglas tubes filled to specification with the same media and gravel as the actual filters in the treatment plant. The pilot filters, which receive the same treated water as the plant filters, are equipped with flow control meters to regulate filtration rates and backwash rates. Surface wash has been added as well as sample taps which provide access to the filtrate for analysis. These tubes are regulated to operate at 2, 3, 4, and 5 gpm/ft$^2$ respectively. In this way, operators can determine which treatment technique and filtration rates are optimum for the plant.

The calculations used to determine the square foot area of a five inch tube, and the subsequent flow rate needed to achieve different filtration rates and wash rates are as follows:

Diameter of five inches is 5 in $\times$ 1 ft/12 in = .4166 ft
Radius = D/2 = .2083 ft.
$\pi R^2 = 3.14 \times .2083' \times .2083' = .13624\ ft^2$

PICTURE 4-4. PILOT FILTRATION APPARATUS AT THE LAKE COUNTY, OHIO, AQUARIUS WATER TREATMENT PLANT. UNITS 1, 2, AND 3 ARE IN OPERATION. UNIT 4 HAS JUST BEEN BACKWASHED AND IS OUT OF OPERATION.

The surface area thus calculated, a table can be constructed showing the flow rates necessary to achieve different filtration rates and wash rates.

**Table I Flow Rates for 5 Inch Diameter Tube**

| *Metering Rate gpm* | *Filtration Rate gpm/ft²* | *Flow Rate gpm* | *Backwash Rate gpm/ft²* |
|---|---|---|---|
| *0.135* | *1.0* | *1.0* | *7.34* |
| *0.272* | *2.0* | *1.3* | *9.54* |
| *0.409* | *3.0* | *1.6* | *11.74* |
| *0.544* | *4.0* | *1.9* | *13.94* |
| *0.682* | *5.0* | *2.2* | *16.14* |
| *0.820* | *6.0* | *2.5* | *18.34* |

The media in these tubes consists of 16.5 inches of 1.0 mm anthracite at top; 9 inches of 0.5 mm silica sand next; and 4.5 inches of .30 mm ilmenite at the bottom. This is supported by a typical gravel underdrain system.

Plant operator studies show that there is no appreciable difference in effluent quality (turbidity) between the 2 gpm/ft$^2$ rate and the 5 gpm/ft$^2$ rate after 72 hours of operation. Particle counting was not performed here. Occasionally, operators do find that a freshly washed filter will pass some coliform at the beginning of a run, but this stops after ripening of the filter. Filters must be washed at 18 gpm/ft$^2$ for proper bed expansion in this system during warm water periods.

## Wash-Water Troughs or Gutters

Fiberglass reinforced plastic, or reinforced concrete wash-water troughs are placed above the filter media, usually at a distance which will allow for at least a 50 percent bed expansion without excessive media loss. Some loss, however, can be expected.

These are arranged so that the horizontal flow of the wash-water flowing into them does not exceed 2.5 to 3.5 feet. In other words, they usually are spaced edge to edge six feet apart. It is very important that the overflow weirs be level and the elevation of each gutter be exactly the same. The bottom of the troughs are kept above the expanded media during wash. A gutter may have a flat bottom and the same cross section area covering its entire length, but they usually have a U- or V-shaped bottom and sometimes the outlet end is deeper than the opposite end.

The rate of wash may be checked as follows:

1. Filter or draw down water in filter to top of wash-water trough.
2. Close wash-water drain.
3. Open wash-water valve.
4. With wash-water valve at normal setting determine the time necessary for wash-water to rise one foot in filter.
   Rise in inches per minute = 12 divided by the time in minutes to rise one foot
   Inches per minute divided by 1.6 = gal/ft$^2$/min

## Cleaning of Filter Bed

To assist the backwash in the process of cleaning the filter bed various types of surface wash, or agitation, have been employed. The first attempts along this line were the mechanical rake, air applied by various means and methods, and later, the stationary surface wash system. The stationary surface wash system consists of a grid system placed directly on top of the sand. Water under high pressure (usually in excess of 50 pounds) is jetted through small openings in the lateral piping. These small water jets dig into the surface of the sand and loosen it so that when the wash-water is applied it breaks through the sand uniformly. The surface wash left on during backwashing also scours and cleans the expanded sand.

A more recent development in the cleaning of filters is the self-propelled revolving agitator (Palmer Filter Bed Agitator), (Patented). This agitator is installed just above the normal level of the bed. Water of 50 psi minimum pressure emitting from nozzles causes the agitator to rotate, and in turn, due to the pressure differential thus created, circulates the filter medium during backwashing thus exposing all particles of medium in the bed to the scouring action of the jets. Recent improvement in the design, with a long curved tip on the rotating arms, has greatly improved the cleaning action in the "corners" beyond the circular path of the arms.

The purpose of the backwash is to maintain the medium in a fluid condition, and to carry off the turbid matter which is ground to a fine and light state by the scouring and impinging of grain upon grain of the medium by the high velocity jets. This method will keep a filter clean under the most severe operating conditions. It is recommended that all filters used for filtering lime-softened water be equipped with some type of surface wash or agitation system.

PICTURE 4-5. PALMER FILTER BED AGITATOR, ALLENTOWN, PA.

Failure to clean the bed properly will lead to many problems, and ultimately, filter breakdown. Mudballs will form and accumulate, causing clogged areas in the filter media. These clogged areas contract with increasing loss of head, and cracks due to shrinkage occur. These cracks allow turbidity particles, algae and bacteria to pass on to the clearwell. Clogged areas also permit channeling of backwash water, which can destroy filter bed continuity.

The AWWA subscribes to three methods of backwashing: upflow water wash without auxiliary scour, upflow water wash with surface wash, and upflow water wash with air scour. Upflow water wash with surface wash is the most prevalent form found in the U.S., but auxiliary air scour is finding its way into more newly built filtration plants.

## Pressure Filters

There are two types of pressure filters, vertical and horizontal. See Pics. 4-7 and 4-8. They usually are built to withstand a pressure of from

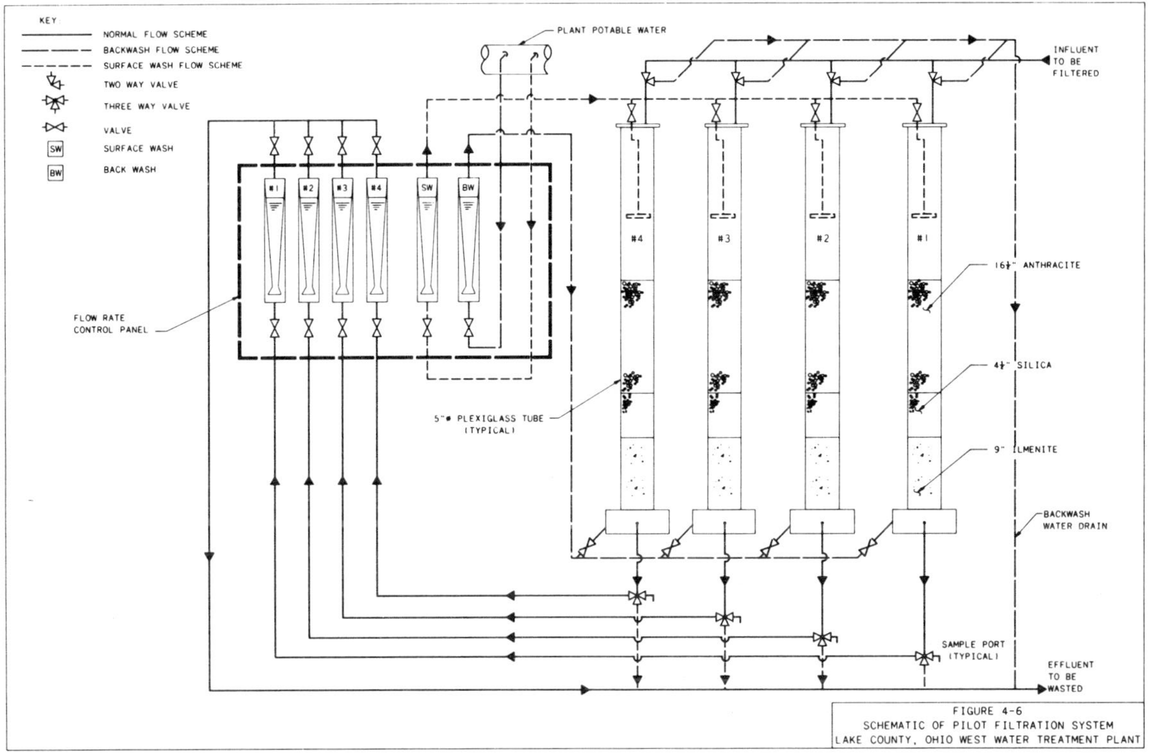


FIGURE 4-6. SCHEMATIC OF THE PILOT FILTRATION ASSEMBLY AT THE LAKE COUNTY, OHIO, AQUARIUS TREATMENT PLANT (COURTESY BURGESS AND NIPLE, LIMITED)

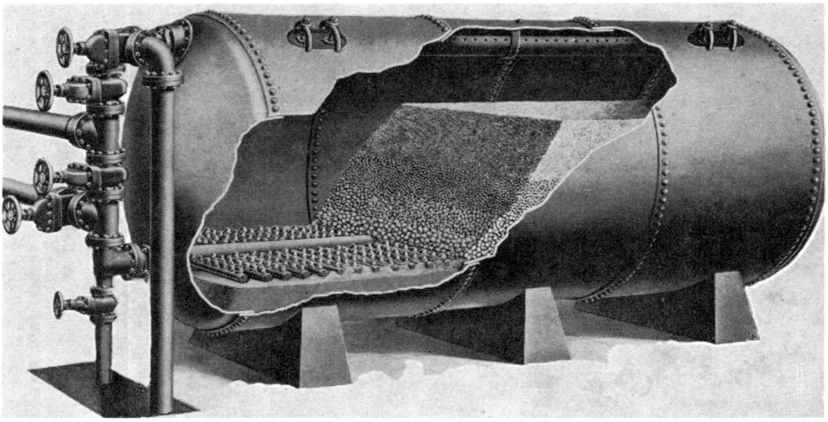

(COURTESY THE PERMUTIT CO.)

PICTURE 4.7. HORIZONTAL TYPE PRESSURE FILTER

50 to 65 pounds per square inch but may be obtained to withstand much higher pressure. Vertical filters range in size from 1 to 10 feet in diameter. Horizontals usually are 10 feet in diameter and about 25 feet in length. They are equipped and operate just about the same as gravity filters except that water is forced through them under pressure.

Although pressure filters have some advantages in that they generally are less expensive to build, are free from air binding, and water under pressure can be passed through them, thus often eliminating

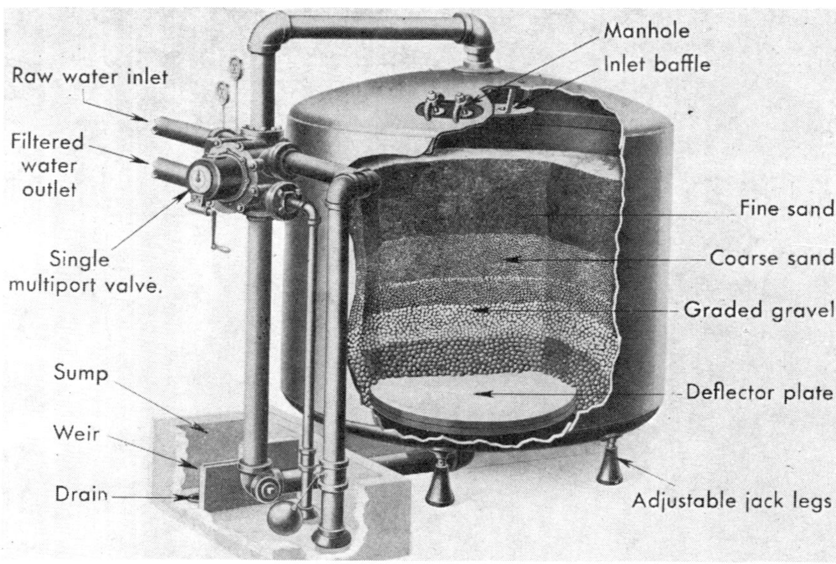


(COURTESY THE PERMUTIT CO.)

PICTURE 4.8. VERTICAL TYPE PRESSURE FILTER

second pumping, they have the disadvantage of preventing easy inspection of the filtering material and observance of the wash-water efficiency. For these last two reasons gravity filters are preferred for filtering municipal water supplies.

All surface water plants must be equipped with some type of filtration. Where well water supplies are to be softened by the lime and soda ash process, it may be more economical to provide longer retention periods (24 to 48 hours) in the mixing tanks and sedimentation basins than it would be to build filters. If the reaction period and settling time is sufficiently long, a clear, stabilized water can be produced without filtration. To insure proper stabilization, a dosage of at least 0.50 mg/L of a polyphosphate salt, such as sodium polyphosphate, should be added to the settled water before it is delivered to the distribution system.

## Diatomaceous Earth Filters

A somewhat different type of filter can be used under certain limiting circumstances. In this method, a layer of diatomaceous earth (skeletons of microscopic animals), held in place on a supporting fine screen, acts as the active filter media. The media is wasted at every filter wash operation and must be replaced at the start of each filter run by (1) an initial "pre-coat", in which a fairly definite amount of the media is added as a slurry before actual start of the filtering operation, and (2) a "body feed" of measured amounts of the material being added along with the water actually filtered. Any interruption of flow during the filtering run will result in the loss of the coating, necessitating the building of a fresh coating before resuming filtration. Increased sludge production and high operating costs are usually associated with earth filters.

These filters can operate under either pressure, vacuum, or both, but require considerably more pressure differential for operation. Coagulants normally will plug the filter so that this type of filter is more apt to be used on water with low turbidity, and microorganism (algae), where coagulation is not required. Their use is generally restricted to small, temporary or portable installations and swimming pools.

Chapter V

# DISINFECTION

The 1986 amendments to the Safe Drinking Water Act provide for four specific regulations which pertain to disinfection in water supplies. Some of these regulations took effect in 1991, and some became effective in June of 1993.

The first regulation changed the traditional way in which we assess distribution system sterility. The standard MPN (most probable number) and Membrane Filter methods for total coliform density have been replaced by a presence/absence concept. This is the total coliform rule (TCR).

The second regulation, the Surface Water Treatment Rule (SWTR), requires the implementation of two treatment techniques in the plant. They are the 99.9% removal or inactivation of Giardia cysts and the 99.99% removal or inactivation of viruses, and the production of a finished water with turbidity levels of less than 0.5 NTU in 95% of samples taken each month.

The third regulation is the Disinfectants/Disinfection By-Products (DBP) Rule. This rule adds new regulated compounds to the TTHM'S, and it sets parameters for disinfectant residuals in the finished water.

The fourth regulation is the Ground Water Disinfection Rule, the major requirement of which is the regulation of viruses in ground water supplies.

All four components of the regulations are designed to optimize disinfection, while keeping a close watch on byproducts which are an unfortunate yet necessary part of the process.

One new concept that has been introduced in water treatment is the CT value. It stands for the Concentration of disinfectant multiplied by the effective contact Time of the reaction basin, and all surface water plants were expected to meet minimum requirements for this value by July of 1993. Water plants must perform a tracer study to determine the effective volume factor of their clear wells, or accept an assigned number for that value which will be given to them by the EPA. The factor, a percentage of one, must be multiplied by the theoretical detention time of a clearwell to determine actual residence time for water in the disinfection process. Clear wells that are baffled, and those that have a greater length to width ratio, will achieve better baffling factors than the old style clear wells. In many cases, plants have had to retrofit old basins, or construct new ones to meet the intent of the regulations.

The units of the CT value are in mg/L– minutes, and an example of the daily calculation is given here:

Calculate the CT value for a basin with a surface area of 115 feet by 75 feet, if the flow through the basin was 4.2 MGD, and the basin water depth was 8 feet. Chlorine residual at the time of the test was 1.6 mg/L, and the effective volume factor of the clearwell is 0.38.

> 4,200,000 gal/Day ÷ 1,440 min/Day = 2917 gpm
> Theoretical Detention Time = 115′ × 75′ × 8′ × 7.48/2917 gpm = 176.9 minutes
> CT = Theoretical detention time x baffling factor × $Cl_2$ residual
> CT = 176.9 minutes × 0.38 × 1.6 mg/L = 107.5 mg/L– minutes.

It is clear that the old concept of building clear wells to a theoretical detention time of 4 hours is no longer sufficient for plants which are treating water at near plant capacity.

## Disinfection with Chlorine

Disinfection of water may be accomplished in a number of ways, but the cheapest, most effective, and most generally used method is chlorination. Water that has been treated efficiently by the processes of coagulation, sedimentation and filtration should be nearly free of harmful bacteria. However, complete removal of pathogenic bacteria is essential, and to accomplish this, chlorine is applied to the filter effluent. In modern practice chlorine is applied also prior to filtration to produce a greater factor of safety, and as a function of the treatment processes.

Use of chlorine for disinfection of drinking water has been the subject of much discussion due to its ability to produce trihalomethanes which are possible carcinogenic substances in treated water supplies. For this reason, the public may believe that chlorination is risky and possibly should be discontinued, because of the suspicion that human life may be in jeopardy by the chlorinated hydrocarbons found in very minute amounts.

Trihalomethanes are formed in water by the reaction of chlorine and residual organic material. Several health studies have indicated an association between chlorinated drinking water and increased mortality from some cancers. For this reason, USEPA has established a maximum contaminant level (MCL) of 0.10 mg/L total trihalomethane (TTHM) as a running quarterly average for public water supplies. As of this writing, USEPA is considering a decrease in this limit. It is likely that water purveyors will be required to produce a water with no more than 0.04 mg/L TTHM by the turn of the century. It is the responsibility of the water purveyor to limit the potential formation of these chlorinated byproducts either through modified disinfection practices or through other treatment means.

Although disinfection can be accomplished by other means, there is no feasible complete substitute for chlorine, at least to replace the residual chlorine increment that provides the vital margin of safety for drinking water. Water disinfection by high pH levels induced by lime is most effective and is used to reduce the total chlorine requirement, but it cannot replace residual chlorine. Plants using lime softening in advance of chlorination should have no worry over chlorine-forming carcinogens, since coincidentally, lime removes most of the organic compounds as it softens the water.

In the early days of chlorination of public water supplies, chloride of lime, or so-called bleaching powder, was used almost exclusively as the chlorinating agent. It is a product made by passing chlorine gas over dry slaked lime and usually contains 33 to 35 per cent available chlorine. Modern manufacturing processes have made it possible to prepare soluble powdered chlorine compounds which have an available chlorine content of from 65 to 70 per cent. A solution of these compounds usually is applied to the water being treated through orifices of suitable equipment. Hypochlorinators, which are small, accurate, diaphragm pumps, most commonly are employed for this purpose.

Chlorine generated in electrolytic cells at the water plant also has been used but is not a method which has been widely adopted.

## Chlorine Gas

Chlorine gas has been found to be more economical and effective in its application than chlorine compounds. Therefore, liquid chlorine supplied in steel cylinders has almost completely supplanted the chlorine compounds for general use. Hypochlorites are still used for specific purposes such as emergency applications, water main disinfection, and in very small water supply systems.

Some of the properties of chlorine gas are as follows:

1. A yellowish-green gas.
2. Has a sharp pungent odor.
3. Is about two and one-half times as heavy as air.
4. Is easily condensed to a liquid.

5. Is inactive to metals when dry and can be stored indefinitely in steel cylinders.
6. Is very corrosive to most common metals when wet. Since it usually is applied dissolved in water, it must be measured and carried through non-corrosive materials such as glass, rubber, silver, and several newly developed rigid and flexible plastics.
7. Is extremely irritating to the membranes in the nose, throat, and lungs. Therefore, great care must be exercised in handling and applying it.

Some precautions which should be observed:

1. A self-contained breathing apparatus, (SCBA), should be available at all times. It should be frequently inspected to see that it is in good working order and that the air tank is full. The unit should be stored near, but not in, the chlorine room. All employees should receive periodic training in its use.
2. The chlorine room should be sealed completely from the rest of the building with entrance from the outside only.
3. All valves should be checked to see that packing glands are tight. When changing chlorine cylinders, discard old gaskets at the unions, and install new ones. All joints should be inspected for leaks using an open bottle of aqua ammonia. A chlorine leak will show up at once producing a white cloud (appearance of smoke) of ammonium chloride particles.
4. An exhaust fan must be provided with inlet at floor level. The control switch, located outside the chlorine room should be turned on several minutes before entering the room to locate a chlorine leak. Since the chlorine fumes are heavier than air, avoid stooping over in hunting leaks.
5. Safety repair kits, as available from companies making chlorine, should be available to repair leaking cylinders.
6. All personnel should be instructed on emergency procedures to follow in case of chlorine leaks; i.e., where the SCBA and exhaust switch are located, where the shut-off valves are (these should be properly labeled) and how to use the repair kits.
7. Heat should never be applied directly to any Chlorine container.

### *Operation of V-Notch Chlorinator*

In the Series A-711 chlorinator (Fig. 5-1), the chlorine gas is metered under a vacuum developed by an aspirator type injector (A). The gas

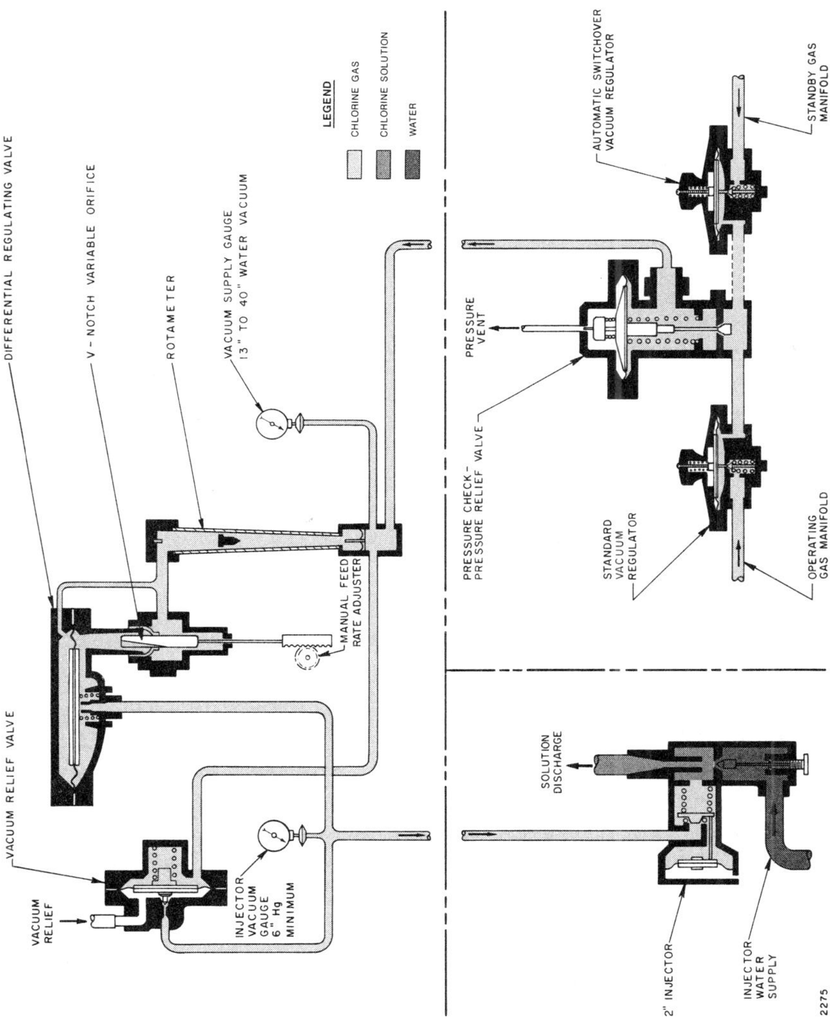


FIGURE 5-1. FLOW DIAGRAM FOR WALLACE AND TIERNAN V-NOTCH CHLORINATOR

enters the chlorinator through a spring-loaded diaphragm operated pressure reducing valve (B). This valve maintains the proper operating vacuum ahead of the V-notch Variable-Orifice.

The gas next flows through a rotameter rate of feed indicator (C) and then through the V-notch Variable-Orifice (D). After leaving the orifice, the gas passes through a vacuum differential valve (E), which maintains a constant differential across the V-notch, and then to the injector where it is dissolved in water. The resultant solution is discharged from the injector to the point of application.

With the vacuum differential across the V-notch Orifice (D) maintained at a constant value under all conditions, the feed rate is adjusted by changing the area of the V-notch plug within its ring.

The chlorine pressure reducing valve, which regulates the vacuum ahead of the metering orifice, also shuts off the chlorine if interruption of the injector water supply should destroy the operating vacuum.

The injector water supply is turned off or on by a three-way pilot valve (F) which controls a spring loaded diaphragm actuator. Thus, starting or stopping the chlorinator is a simple matter of positioning the three-way pilot valve.

Intermittent start-stop or program operation is obtained by interrupting the injector water supply. The injector shut-off pilot valve is equipped with a three-way solenoid valve which functions in place of the manual pilot valve to start or stop the chlorinator in synchronism with a program timer or other equipment operating circuit.

Automatic operation is accomplished by positioning the V-notch plug to change the size of the orifice and thereby adjust the feed rate. The operation of auxiliary equipment for automatically positioning the plug is described in other catalog files.

*Operation of Fischer Porter Chlorinator*

Chlorine gas enters the chlorinator (Fig. 5-2) at the inlet connection block, where it is mechanically filtered to remove foreign matter and heated to evaporate any liquefied chlorine. The gas then enters the vacuum regulator where its pressure is reduced to a vacuum. The gas is then metered in the Flowrator meter. The gas passes through the rate valve, which dictates the flow of chlorine gas.

The gas then enters the differential pressure regulator, which maintains a constant pressure drop across the rate valve. The gas then passes through the stack isolating valve to the ejector, where it mixes with the water to form a chlorine solution.

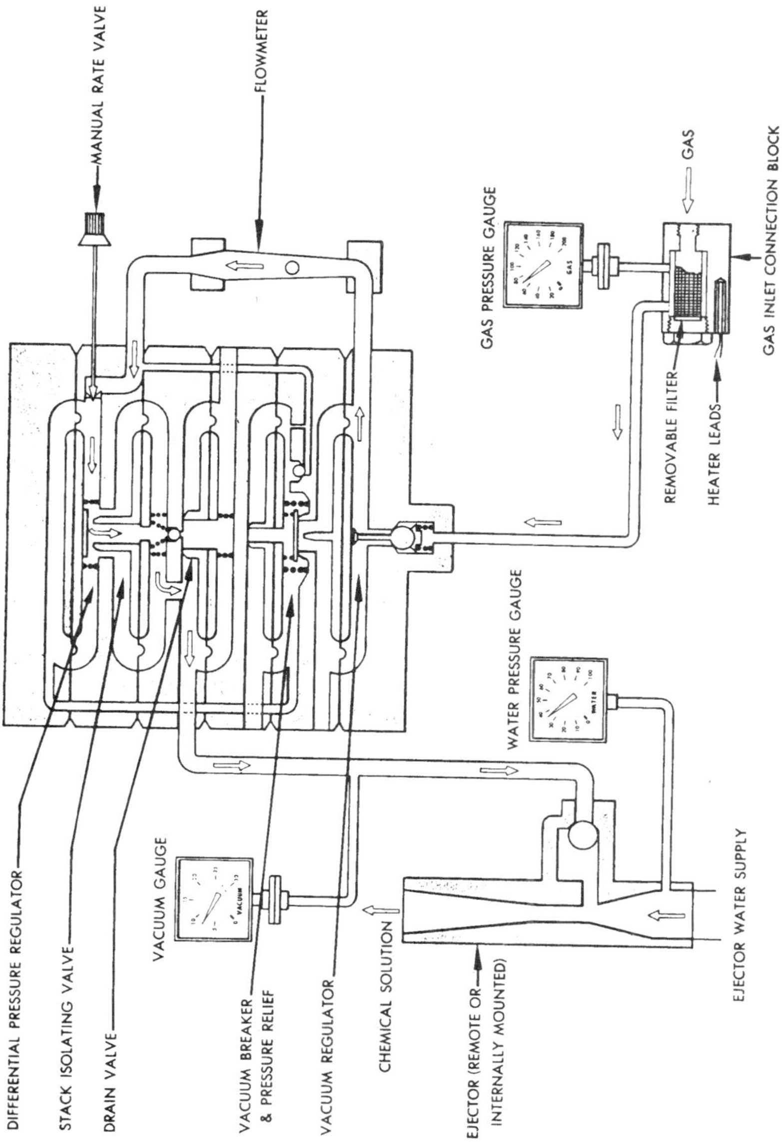


FIGURE 5-2. FISCHER AND PORTER CHLORINATOR

If the water supply is interrupted and/or the vacuum lost, the vacuum regulator diaphragm automatically closes the chlorine inlet supply valve. A water back-check valve is supplied to prevent water feed-back into the stack regulator from the ejector. Should this back-check valve leak, the stack water relief valve will permit water to drain from the chlorinator.

If the vacuum becomes excessive, the vacuum relief valve admits air into the control stack. However, the air bypasses the flow meter and is not metered as chlorine gas. If the system becomes over pressured, a pressure relief valve vents the system to the atmosphere.

### *Operation of BIF Chlorinizer*

Chlorine at supply pressure enters the inlet valve (Fig. 5-3), passes through a ceramic filter and is stopped by ball (1), which is held against its seat by a compressed spring. When the injector water supply is turned on, water flows through the eductor section of the injector

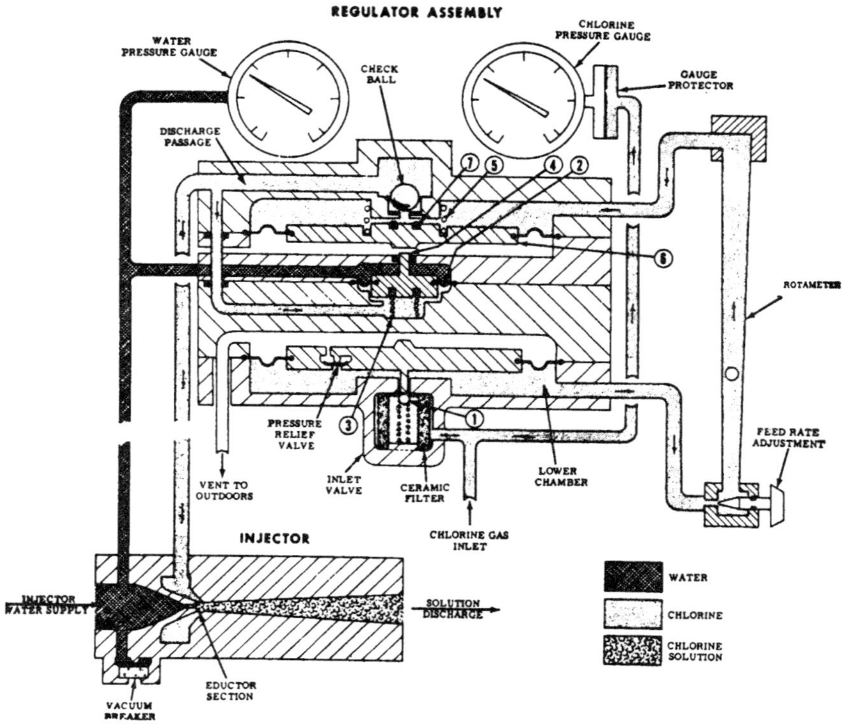


FIGURE 5-3. BIF CHLORINIZER FLOW DIAGRAM

creating a vacuum. The Vacuum acts on the lower side of the small diaphragm (2) and the discharge passage of the regulator assembly.

When the differential force caused by water acting on the upper surface of the small diaphragm and vacuum acting on the lower side is great enough to overcome the force of spring (3), pin and diaphragm (4) move down. Spring (5) will now push down upper diaphragm plate (6). The difference between pressure acting on the bottom of the check ball and pressure acting on the top causes the check ball to rise. The inside of the regulator is now under a vacuum. When pressure on both sides of the check ball is equal, the check ball seats.

Vacuum in the lower chamber pulls the lower diaphragm plate down, pushes ball (1) off its seat and admits chlorine gas into lower chamber. Chlorine gas flows through the feed rate adjustment and rotameter into area above upper diaphragm. Spring (5) holds upper diaphragm down. The flow of chlorine unseats the check ball and passes through to the eductor section of the injector where it is mixed with injector supply water and discharged as chlorine solution. When the injector water supply is shut off, there is equal pressure on both sides of small diaphragm. Compression spring (3), stronger than spring (5), pushes the small diaphragm upwards. Pin (4) pushes against the upper diaphragm plate sealing "O" ring (7) against the top section. This prevents water or water vapor in the injector suction tubing from entering the control section of the regulator assembly.

For injector back pressures from 50 to 130 psig, the hydraulically-operated small diaphragm is eliminated, and a float operated check valve with resilient "O" ring seal is mounted on the injector.

A pressure relief valve, in the lower diaphragm is held closed by vacuum. If lower chamber is pressurized, the relief valve opens and chlorine gas vents to outside of building. A vacuum breaker, in the injector, admits air into water line when injector supply is shut off and a vacuum exists in the solution discharge line.

### *Operation of Advance Gas Chlorinator*

In the Gas Chlorinator system (Fig. 5-4), water flowing continuously (or intermittently for start-stop operation) through a Venturi (1) creates operating vacuum. The vacuum causes the spring-loaded check valve (2) to open, and vacuum is impressed on the regulating diaphragm assembly (3). Chlorine gas cylinder pressure is reduced to a regulated vacuum pressure at the inlet valve (4). The rate of chlorine gas feed is indicated directly in pounds per 24 hours on the rate indicator (5), and the quantity of chlorine allowed to flow is adjusted with the rate valve (6) (See Fig. 5-4 on page 89).

## How Chlorine Kills Bacteria

The effect of chlorine on organisms in water and sewage has been the subject of considerable controversy. It was believed originally that nascent oxygen, produced by the reaction of chlorine with water, accomplished the destruction of bacteria. The following reactions illustrate what is thought to take place:

$$H_2O + Cl_2 \rightarrow HCl + HOCl$$
$$HOCl \rightarrow HCl + O$$

Explanation of reaction: Chlorine unites with water to form hypochlorous acid which, being unstable, breaks up into nascent oxygen and hydrochloric acid. It may be true that some of the organisms are destroyed in this manner. It is known, however, that potassium permanganate, which is a more powerful oxidizing agent than chlorine, is less effective as a germicide. It is know also that when chlorine and ammonia are combined to form chloramine ($Cl_2 + NH_3 \rightarrow NH_2Cl + HCl$), the resulting compound has little if any oxidizing power, but still is very effective as a disinfectant. Hence, a further reason than oxidation alone must be looked for to explain the disinfecting power of chlorine. A plausible explanation is the theory that chlorine unites, at least in part, with the cell structure of the organism to form chloro-products that act as toxic poisons to these organisms. In fact, the bodies of some bacteria completely disappear as the result of chlorination—probably converted into soluble compounds.

## Amount of Chlorine Required

It frequently is impossible to predict accurately how much chlorine must be added to accomplish satisfactory disinfection, as some waters will absorb much more chlorine than others. For example, the presence of reducing substances such as organic matter, sulfides, and unoxidized iron and manganese, will cause the absorption of chlorine. Tests for the presence of "free" chlorine in the water five minutes after the chlorine application will indicate whether the proper amount was applied initially. The slight amount of chlorine in excess of the demand is called residual chlorine, and represents the difference between that originally applied and that absorbed by the water.

It is obvious that a badly polluted river water containing organic matter and other impurities will absorb more chlorine (have a higher chlorine demand) than a comparatively pure well or spring water. Residual

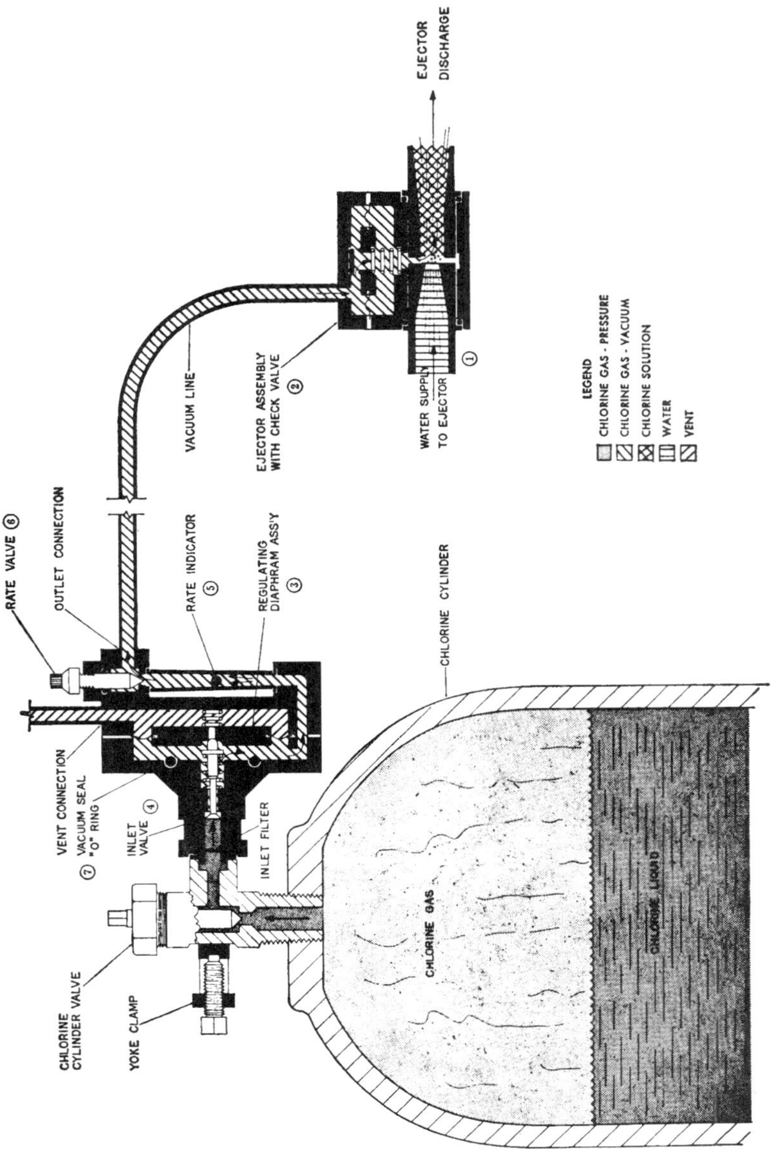


FIGURE 5-4. ADVANCE GAS CHLORINATOR FLOW DIAGRAM

PICTURE 5-1. AMMONIA FEEDERS AND BULK STORAGE TANKS AT THE MAHONING VALLEY SANITARY DISTRICT

chlorine represents a factor of safety and indicates that all organisms have been destroyed. This residual chlorine usually is measured amperometrically, or colorimetrically with DPD (N-N-Diethyl-p-phenylene) solution. (DPD and chlorine produce a pink or red color.)

## Ammonia-Chlorine

In cases where comparatively high residuals are necessary to produce safe water, or where a reaction between phenols and applied chlorine produces chlorophenol tastes and odors, various means have been used to prevent these problems. One very effective method is the use of ammonia. This can be applied as a compound or as a gas. Ammonia gas usually is obtained in cylinders and applied through measuring equipment similar to chlorinators. Chlorine and ammonia are used separately, and a new sterilizing compound called chloramine is formed in the water.

The following reactions occur:

$$2Cl_2 + NH_3 \rightarrow NHCl_2 + 2HCl, \text{ or}$$
$$Cl_2 + NH_3 \rightarrow NH_2Cl + HCl$$

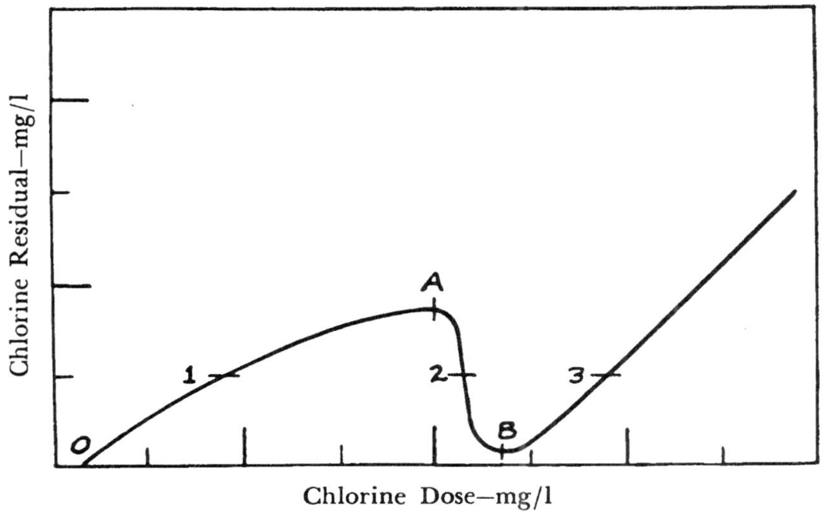


FIGURE 5-5. "BREAK-POINT" CHLORINATION CURVE

The first is dichloramine and the second, monochloramine. Chloramines are relatively tasteless and enable a much higher chlorine residual to be carried without producing chlorinous taste or odor problems. The optimum ratio of chlorine and ammonia generally lies between four and eight parts of chlorine to one part of ammonia.

The death rate of bacteria produced with chloramines is considered to be somewhat slower than that obtained with chlorine alone. Length of contact period between treatment and consumption should determine the points of application and the quantities to be used.

## Break-Point Chlorination

A widely used technique in the control of chlorination has been instituted which affords a more successful use of chlorine for controlling tastes, odors, and bacteria. The term "break-point" chlorination is used in connection with this practice and is explained as follows:

On the addition of increasing doses of chlorine to a series of water samples the residual chlorine gradually increases to a point where further additions of chlorine are followed by a decrease in the residual. Finally on still further additions of chlorine the residuals again begin to increase. This second point is designated as the "break-point" and is the point at which free residual chlorine begins to develop.

In many instances tastes and odors have been minimized or eliminated entirely when sufficient chlorine has been added to produce a break-point in the residual.

The "break-point" chlorination process in many cases affords better results in control of tastes and odors and removal of bacteria. The break-point for any given water supply usually is obtained experimentally by adding increments of chlorine to a series of samples, mixing, and determining the total chlorine residual after suitable contact time, usually 30 minutes. A study of the curve obtained on plotting dosage against residual (Fig. 5-5) may look somewhat like this:

In the region O to A the chlorine is in a combined form, usually as chloramine, and the taste and odor concentration may be increased with dosage increase. At point A sufficient chlorine potential is present to break apart the combined form and at point B this reaction is complete, most of the chlorine being consumed and no longer showing any appreciable residual. From point B, the break-point, an increase in dosage will increase the residual in the same amount, and the chlorine is present as free chlorine.

It will be seen that there are three points, representing three dosages, where there is the same total chlorine residual. Point 1 is in the region of possible taste and odor difficulties, point 2 is in the region of chlorine utilization and should be avoided, since it is on the steep part of the curve, where a slight variation in either chlorine demand or chlorine dose may shift the residual to point A, or extreme tastes and odors, to point B, with insufficient chlorine residual. For optimum results, where "break- point" is used, point 3 represents desirable dosage. The free chlorine, represented by the total residual at point 3, is found to be many times more effective in bacteria killing power, than the same total residual represented by point 1 or 2.

## Dechlorination

The quantity of chlorine remaining in a treated water may be within the desired range. If it is not, a chemical dechlorinating treatment may be required to obtain the desired final residual. This can be accomplished by the addition of sulfur dioxide, sodium thiosulfate, sodium bisulfate, sodium sulfite, or activated carbon.

## Chlorine Dioxide ($ClO_2$)

Chlorine Dioxide ($ClO_2$) is even more effective than chlorine in disinfecting water and destroying certain tastes and odors, particularly those caused by phenolic wastes. Because the cost is greater than chlorine, the chlorine dioxide application generally is a final step, after other chemicals, including pre-chlorine, have been applied. However, $ClO_2$ is finding increased use as a disinfectant because it does not create trihalomethanes like $Cl_2$ does. Therefore, the additional cost may be offset by the need for change in disinfection strategy.

Chlorine dioxide usually is generated at the water plant by the reaction between sodium chlorite and chlorine. The reaction $Cl_2 + 2NaClO_2 \rightarrow 2ClO_2 + 2NaCl$ requires a theoretical ratio of 1 lb. of chlorine to 2.63 lb. of sodium chlorite. However, a large excess of chlorine is used (about 500 mg/l of available chlorine) to drive the reaction to completion and to provide the necessary pH level (pH 3.5 or less) for the reaction. If chlorine in the form of sodium hypochlorite is to be used, it must be acidified for the reaction to proceed.

Great care must be taken with this chemical as it reacts violently with reducing agents. It has also been reported that gaseous $ClO_2$ is explosive in concentrations as low as 10% in air.

In actual practice, a solution of sodium chlorite, made up in a separate container, is fed by means of a metering pump into mixing chamber or generator through which the strong chlorine solution is passing. This resulting solution, a mixture of $Cl_2$ and $ClO_2$, is usually applied to the filtered water at the normal point of post-chlorination.

More and more plants treating surface supplies have installed facilities for chlorine dioxide even though it may be used only part of the time. See Chapter VII on Tastes and Odors for further reference to $ClO_2$. For laboratory procedures to determine break-point chlorination or chlorine dioxide dosages, see Chapter XVIII.

## Ozone

When dry air or oxygen is passed through a gap between two highly charged electrodes, some of the oxygen is converted to ozone, $O_3$. This highly unstable colorless gas reacts with almost anything in water, including bacteria, which it kills. The half-life of ozone in water (the time it takes for half of the concentration of ozone to dissipate), is calculated to be about one minute; therefore, ozone is usually used as a disinfectant in the pre-treatment scheme. With such a short half-life, the maintenance of an ozone residual in the distribution system is quite difficult.

Ozone finds its place in water treatment where there is concern for trihalomethane formation from Chlorine. Ozone destroys the humic substances which act as precursors for the TTHM's. Ozone is also employed in the removal and control of tastes and odors, and will be discussed further in Chapter VII.

## Disinfection with Lime

The late Dr. A. C. Houston, chemist of the Metropolitan Water Board, London, England, first proposed the excess lime method of sterilizing water*. His results were first confirmed by Hoover and Scott at Columbus, Ohio**, and have since been confirmed at a number of water softening plants. The following table shows the average B. coli indices on raw surface and lime-softened, filtered, but unchlorinated water in six Ohio municipalities:

| | *B.Coli/100 mls.* | | *Hours Detention Period* | | | |
|---|---|---|---|---|---|---|
| | *Raw Water* | *Lime Softened Filtered Water* | *Mixing* | *Clarifier* | *Sedimentation* | *Total* |
| Columbus | 402 | 0.40 | 1.0 | — | 10.0 | 11.0 |
| Fremont | 3,274 | 1.00 | 1.5 | 3.0 | 6.0 | 10.5 |
| Greenville | 1,537 | 1.40 | 1.6 | 0.8 | 6.7 | 9.1 |
| Newark | 1,376 | 0.40 | 2.5 | 0.75 | 20.0 | 23.2 |
| Piqua | 1,004 | 0.30 | 1.5 | 4.0 | 6.0 | 11.5 |
| Youngstown | 6,052 | 0.05 | — | — | 15.0 | 15.0 |

The results showed that all of these plants produced water that met the U.S. Treasury Department standards without the use of chlorine. The average load of all the plants was 2274 B. coli per 100 milliliters, whereas the treated water averaged only 0.6 B. coli per 100 milliliters. Some of the plants named in the table did not use chlorine at all. Some used it all the time, and others only occasionally. Where it was used as an additional factor of safety it was applied after filtration. The plants

* Eighth Research Report of the Metropolitan Water Board, London, England.

** Engineering Record, September 6, 1913, page 257.

studied had detention periods, on design bases, varying from 3.5 to 9.1 hours. However, they were not operated at full capacity, therefore the detention periods shown in the table are based on 1929 average consumption records, and vary from 9.1 to 23.2 hours. On the other hand, some of the plants did operate at full capacity during a portion of the day, and then shut down, so that the period of contact may be said to have ranged from 3.5 to 23.2 hours.

The important factors in lime disinfection are:

1. Quantity of excess lime used.
2. Time or reaction.
3. Quantity of mud, organic matter, and bacteria present in raw water.

Numerous Ohio well supplies, previously considered dangerous because of leaky piping and casings, have been made safe by disinfection obtained simultaneously from conventional lime softening. In several such small plants no chlorine was used. Disinfection was obtained solely with lime. Today such plants cannot operate this way since the Ohio EPA regulations require chlorination of all public water supplies whether bacteria is present or not.

## Other Conclusions on Bactericidal Action of Lime

1. Houston concludes that quicklime added to raw Thames River water in the proportion of 1 part of quicklime to 5000 parts of water (about 11.7 grains per gallon) kills B. coli in 5 to 24 hours.
2. At water plants in Ohio using enough lime to produce a water with a pH value greater than 9.5, B. coli were not found in 50-milliliter portions out of a total of 25 samples tested, whereas at those plants where only enough lime was added to assist in coagulation (pH less than 9.5) B. coli were present, with one exception, in 22 samples in 50-milliliter and usually in 10-milliliter portions.
3. At Columbus, Ohio, when enough lime is added to Scioto River water, containing 300,000 B. coli per 100 milliliters, to absorb the free and half-bound carbonic acid, and to precipitate the magnesium, the bacteria of the colon and typhoid group are reduced 99.67 per cent in 5 hours after being so treated. The

germicidal action is 99.93 per cent effective in 5 hours when an excess of approximately one grain per gallon is added. The treatment is 100 per cent effective in 5 hours when 3 grains excess are used. When organic matter and the B. coli index are low, a shorter reaction time is necessary than when organic matter and the B. coli index are high.

4. Results of laboratory experiments show that lime softened water inoculated with typhoid organisms soon becomes free from them.

## 1968–75 Lime Disinfection Research

Research by the EPA and other agencies, mainly in the treatment of wastewater, confirms the above earlier findings on the disinfection action of lime and also expands this knowledge.

Pathogenic bacteria survival is adversely affected by high pH, elevated temperatures, and time of exposure. Evidence that the survival rate of bacteria decreases with increasing (lime-induced) pH levels is indicated in the accompanying graph, based on research by Grabow on the varying survival of different strains of bacteria at a pH of 11.0 with lime treatment. Most bacteria specimens were decimated in durations of 60 to 120 min. But with three strains of bacteria at 120 min. exposure only about 70 to 95% of the bacteria was killed.

While it has been found that some types of bacteria can be removed at a pH as low as 9.5, other strains may require a pH of over 11.0, such as 11.5 or even 12.0. Generally the higher the pH the less contact time is needed, or longer contact times can produce comparable bacterial reduction at slightly lower pH levels. Bacteria are more resistive at low temperatures, i.e., 1 to 15°C, but Morrison and Martin found that rapid destruction of coliform bacteria occurred at pH levels of 11.5 to 12.0 even at 1°C temperature, whereas pH levels just below 11.0 were ineffective. Doyle found that salmonella typhosa bacteria that causes typhoid fever, is destroyed at pH 11 after 2 hours.

Increasing interest is evident in the effect of high pH on another pathogen—VIRUS—since chlorine has many limitations in virus elimination. Thus, high alkalinity treatment with lime may prove to be a valuable complement to chlorine in effecting greater public protection from diseases. Lime is devastating to poliovirus I, an example of which is taken from research by Berg et al, using lime flocculation and sand filtration as follows:

| | Lime Treatment | |
|---|---|---|
| *pH* | *Virus Remaining pfu/L (a)* | *% Virus Removed* |
| 9.27 | 336 | 99.0 |
| 10.13 | 685 | 98.7 |
| 10.88 | 1.7 | 99.995 |
| 11.10 | 5.1 | 99.98 |
| 11.2 | <1.4 | >99.997 |

(a) Concentration of virus expressed as plaque-forming units per volume or pfu/L. Initial virus concentration was 33,000 to 55,000 pfu/l.

Other researchers have indications that high pH may be nearly as effective as with polio in eradicating virus from water that causes hepatitis and other diseases, as well as some parasite eggs from worms in fecal matter. While the eggs flocculate readily with lime or other

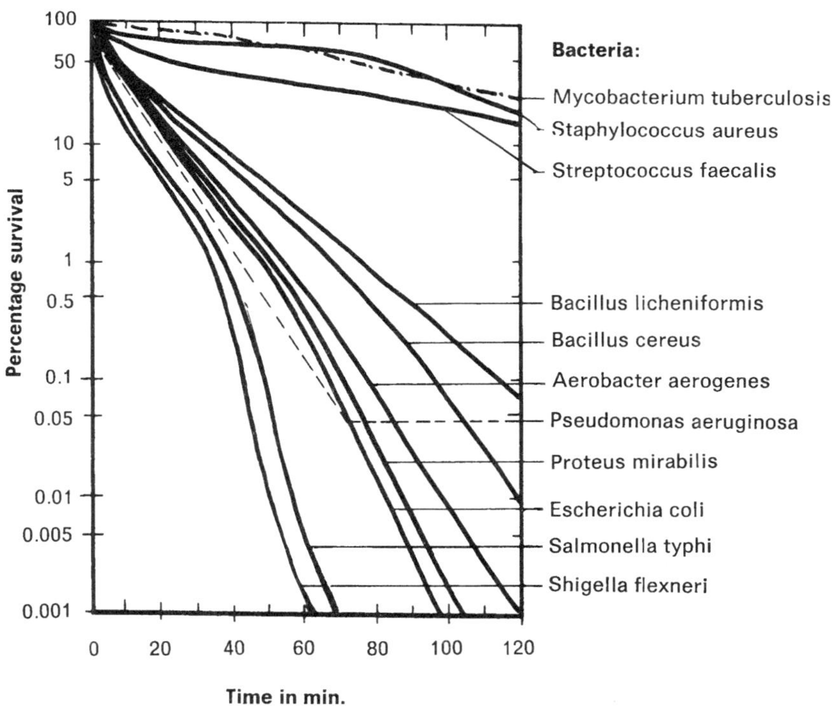


GRAPH 5-1.

coagulants and settle with other sediment, the high pH destroys those parasite eggs with thin shells after a short duration. However, some eggs with thick shells will resist even the highest pH for up to several weeks.

In all of this bacteria, virus, parasite, and parasite egg removal with lime treatment, and unlike treatment with other coagulants and chlorine disinfection that functions in one manner, lime attacks these pathogens in two ways:

(1) It removes pathogens by physical absorption on the lime particles by forming flocs that settle and (2) by creating a high pH that simply destroys the pathogen.

### *Additional Research*

"Survival of Pure Cultures suspended in Membrane Filtered Humus Tank Effluent adjusted to pH 11 with Lime". Additional research on the effects of lime on poliovirus has been performed by Sattar et al. Their work, published in 1974 in the Journal of Social and Occupational Medicine is quoted here: "The lime (calcium hydroxide) treatment of sewage shows promise in this respect (1) Morrison et al. (2) and R. Brouzes (unpublished data, Domtar Research Centre, Senneville, Quebec) have already shown that addition of enough lime to sewage to achieve a pH value of 11.5 gives as clear supernatant with a low content of bacteria, phosphates, heavy metals and oxidizable materials; the sludge from this process is odorless and also contains few viable bacteria. However, little information is available on the removal and inactivation of viruses during this process. Investigation of the virus destroying efficiency of this method was the specific objective of the present study."

A one litre sample of domestic sewage was experimentally-contaminated with poliovirus 1(Sabin). After a thorough mixing, a 20 ml amount was withdrawn from the sample to act as virus control (fraction #1). To the remaining sample an aqueous slurry of lime was added drop wise and with constant stirring till a pH value of 11.5 was achieved. The sample was stirred for an additional 15 minutes before it was transferred to an Imhofff cone and the sludge was allowed to settle for one hour. To detect the presence of the virus in the resulting supernatant, a 20 ml amount was removed (fraction #2) and its pH was immediately adjusted to 7.2 with 0.1 N HCl. The sludge was drained from the bottom of the cone and centrifuged (5000 g. 20 min). The supernatant was discarded and the sediment was resuspended in 20 ml of 10% fetal calf serum in normal saline. The pH of the suspension was brought to 7.2 (fraction #3). Since calf serum was found highly satisfactory in the elution of virus from particulate materials (3) it was employed here in an attempt to desorb the virus from the sludge.

Although the amount of virus used in these experiments was many times higher than one would expect to find in domestic sewage, these data clearly indicate that during lime treatment of one liter of sewage more than 99.99% of the input poliovirus is rendered non-infectious. These findings warrant further investigations with other groups of viruses known to occur in sewage.

Chapter VI

# Fluoridation of Water

A report published by J. M. Dunning in 1962 showed the relationship of elevated fluoride levels in children to the incidence of dental caries.[1]. This data, and data from subsequent reports shows a minimum incidence of caries when water fluoride levels were held continuously at about 1.0 mg/L. Because of this fact, public water supplies are usually fluoridated to optimum levels. In 1985, the center for disease control stated that 130 million Americans, or 61% of those served by public water supplies, consume fluoridated water. In 1968, this figure was about 80 million.

Dental caries, the most prevalent chronic disease of human beings, is most widespread in children. Tooth decay reaches a peak in adolescence, and diminishes in adulthood. It is for this reason that fluoridated water has the greatest effect on the teeth during the formative years. Fluoride is absorbed on the teeth and bones. The softer tissues of the body do not seem to retain fluoride. When teeth are forming, an adequate source of fluoride will help to strengthen the enamel, forming a more impervious layer to the bacteria which cause decay. Caution is needed, though, in that increased levels of fluoride above the optimum can cause mottling of teeth from dental fluorosis. This condition creates brittleness in the teeth, and moderate to severe discoloration. Water purveyors must take every precaution to control the levels of fluoride ion concentration in their supply. The USEPA has set a maximum contaminant level (MCL) for fluoride at 4 mg/L to prevent crippling skeletal fluorosis.

A secondary MCL of 2 mg/L was established by USEPA to prevent dental fluorosis. These limits are reviewed every three years. It is generally agreed that a fluoride level of 0.8 to 1.3 mg/L Fluoride is most effective, dependent on temperature. (In hotter regions, people tend to drink more water, and therefore will retain more fluoride.)

---

1. Principles of Dental Public Health, Harvard University Press, 1962.

## Chemicals Used in Fluoridation

Fluoride is the negative ionic form of Fluorine, a gaseous element. Very abundant in nature, fluorine can be found in all soils, water supplies, plants and animals in varying amounts. Fluorine is a reactive element and therefore, it is always found in combination form, never in the free state. Fluoride-containing minerals are used by industry for fluoride compound production, the most prevalent being fluorspar, cryolite and apatite. From these minerals we get the three most common fluoride adjustment compounds: Sodium Fluoride (NaF); hydrofluosilicic acid ($H_2SiF_6$); and Sodium Silicofluoride ($Na_2SiF_6$). These three compounds are by- products of the phosphoric acid fertilizer industry, and the availability of them is dictated by agricultural production. Sales of fertilizer will have a direct effect on the manufacture and availability of fluorides for water treatment.

## Sodium Fluoride

Sodium Fluoride (NaF) is a white, odorless powder or crystal. It has a specific gravity of 2.79, a molecular weight of 42 (Sodium = 23, Fluorine = 19), and a solubility of 4 grams/100 mls, or 4% in water. When added to water, the NaF will dissociate into Sodium and the needed Fluoride ions. NaF can be purchased in quantities which are very pure, usually near 97 to 98%.

A dosage of 19 lbs. NaF into 1 million gallons of Fluoride free water will produce a 1 mg/L $F^-$ solution. The percent of available Fluoride in NaF is 45% (19 ÷ 42).

Calculation for an estimated dosage of 1mg/L $F^-$ when using NaF (97% purity):

19 lbs / MG NaF × (1 mg/L / 8.34 lbs/MG) × 0.97 purity = 1.0 mg/L $F^-$

## Sodium Silicofluoride

Sodium Silicofluoride ($Na_2SiF_6$) is a white odorless crystalline material. It has a specific gravity of 2.679, a molecular weight of 188 (2 Sodiums = 46, S1 = 28, 6 Fluorines = 114), and a solubility ranging from 0.44 to about 2.5%. $Na_2SiF_6$ can be purchased in purities of 98% or higher.

A dosage of 14 lbs $Na_2SiF_6$ into 1 million gallons of Fluoride free water will produce a 1 mg/l $F^-$ solution. The percent available Fluoride in $Na_2SiF_6$ is about 61% (114 ÷ 188).

Calculation for an estimated dosage of 1 mg/L $F^-$ when using $Na_2SiF_6$ (98% purity):

> 14 lbs / MG $Na_2SiF_6$ × (1 mg/L /8.34 lbs/MG) × .61 available $F^-$ × .98 purity = 1.0 mg/L $F^-$

## Hydrofluosilicic Acid

Hydrofluosilicic Acid ($H_2SiF_6$) is a straw or amber colored, transparent, fuming, corrosive liquid. It has a molecular weight of about 144 (2 Hydrogen = 2, Silicon = 28, 6 Fluorines = 114), and can be purchased as a liquid in drums or bulk at 23 to 35 percent.

A dosage of 46 lbs of 23% ($H_2SiF_6$) into 1 Million gallons of Fluoride free water will produce a 1 mg/L $F^-$ solution. The percent of available Fluoride in ($H_2SiF_6$) is about 79% (114 ÷ 144).

> Calculation for an estimated dosage of 1 mg/L $F^-$ when using ($H_2SiF_6$) (23% purity):
> 46 lbs ($H_2SiF_6$) × 1 mg/L × .79 available $F^-$ × .23 Purity 8.34 lb/MG = 1.0 mg/L $F^-$

It can be noted that water has no "fluoride demand", and therefore a contact period such as for chlorine is not needed. Some fluoride ions will, however, be lost in filtration (usually 0.1 to 0.3 mg/L), so the optimum feed point for Fluoride is in the post filtration scheme. It is prudent to perform daily checks of the Fluoride concentration on plant tap water, and to perform hourly checks of the feed system while in operation.

## Feed Equipment

There are 3 accepted methods for feeding supplemental fluoride to the supply:

1. The Fluoride saturator
2. Dry Chemical Feeder
3. Chemical solution feeder, using either ($H_2SiF_6$) acid or a solution made from the dry chemicals.

The selection of feed equipment is based on many factors, including size of plant, availability and type of fluoride source, cost, and technical expertise of personnel. The larger facilities seem to prefer the liquid storage and feed systems. The operator is directed to the publication Water Fluoridation Principles and Practices, AWWA M4 by the American Water Works Association.

Fluoride chemicals are added to water as liquids, but may be measured as either liquid or solid. Solid chemicals are dissolved into solution before feeding. If so, the strength of the solution must be controlled, usually by feeding a predetermined chemical for a specific duration. Two types of dry-feed machines exist: the volumetric and the gravimetric. The volumetric delivers a predetermined volume of dry chemical within a given time, and the gravimetric delivers a predetermined weight of chemical within a given time, and is more accurate. Figures 6-1 and 6-2 are examples of modern day fluoride feeders.

## Fluoride Removal with Lime

Defluoridation, or removal of excess fluoride is necessary where the source of supply contains levels which are detrimental to the health and welfare of the public. Fluoride can be removed in the softening process where waters are high in magnesium content. Lime will precipitate Magnesium compounds at high pH levels. At these levels, magnesium fluoride precipitates just as does magnesium hydroxide. The sludge is settled in the sedimentation basins and discarded. This discovery was published by R. D. Scott in 1937 in the Journal AWWA in his work titles "Fluoride in Ohio Water Supplies".

Ion exchange and adsorption on bone char and activated alumina are other methods used in the removal of fluorides.

Activate alumina is a porous inorganic adsorbent similar to the ion-exchange resins. It works well because the fluoride ion is "preferred" by the alumina over most other ions. This process is finding increased use in the home as a point of entry (POE) or point of use (POU) device.

## Safety Precautions

Fluoride dusts from such chemicals as NaF and $Na_2SiF_6$ can expose operators to higher than optimal levels of fluoride, and can cause irritation of the nasal and mucous membranes. In addition, fluoride acids can burn the skin. The following precautions are advisable when handling these chemicals:

1. Avoid breathing fluoride dusts. Dust masks should be issued to all those who work with the fluoride dusts.
2. Wear rubber gloves and aprons when handling dusts.
3. Wash away any dusts from the skin immediately.
4. When handling hydrofluosilicic acid, one should wear acid proof gloves, aprons and face shields. Wash these items after use.
5. Do not allow persons with open cuts or sores to handle any fluoride acids or dusts.

SERIES 47–050 SATURATOR

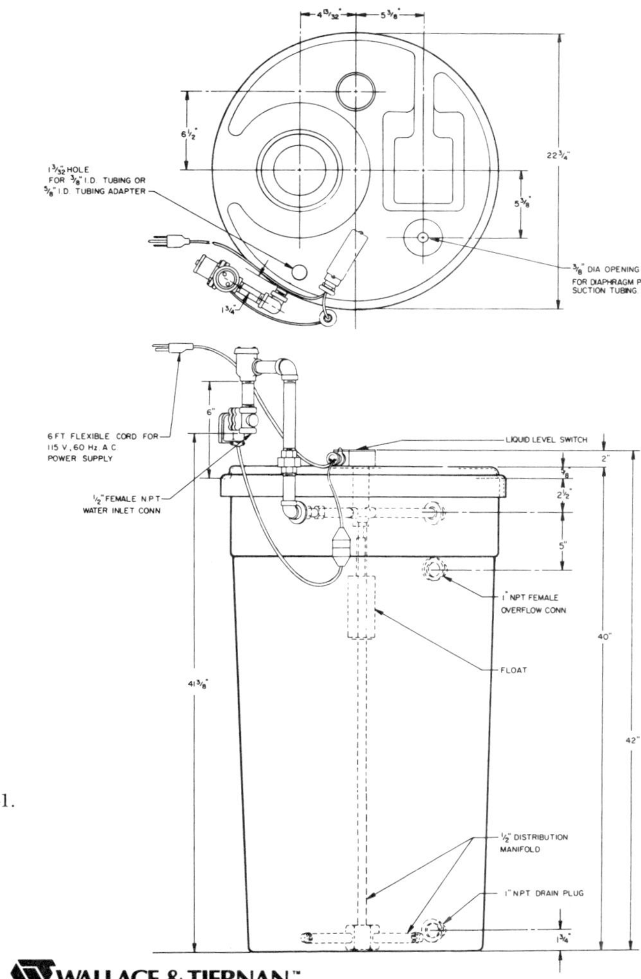


FIGURE 6-1.

WALLACE & TIERNAN™
WALLACE & TIERNAN, INC.
25 MAIN ST., BELLEVILLE, N.J. 07109-3057

6. Be sure to post safety rules in the vicinity of the chemical storage and handling area.
7. Fluoride containers should be labeled as poisonous, and should never be reused for other purposes.
8. If accidental ingestion occurs, give the operator a 1% solution of calcium chloride, a glass of saturated lime ($Ca(OH)_2$) water, or milk to drink. Flush the chemical out of the eyes with warm water.

Get medical attention immediately.

## Sample Dosage Scenario

A water plant treats 7.21 MGD, and wishes to keep a Fluoride ion level of 1.1 mg/L in the plant effluent. If the raw water source contains 0.36 mg/L natural Fluoride ion, how many gallons of 24.3% $H_2SiF_6$ will be needed in the month of September? Acid is 10.5 lb/gallon.

Answer: $H_2SiF_6$ has 79% available Fluoride ion
1.1 mg/L Desired – 0.36 mg/L available = 0.74 mg/L $F^-$ needed dosage
0.74 mg/L $F^-$ × (8.34 lb/MG $F^-$ / 1 mg/L) × 7.21 MGD × (30 Days / month) = 1334.92 lb $F^-$/month
1334.92 lb $F^-$/month ÷ (.243 × .79) = 6953.8 lb $H_2SiF_6$/month
(6953.8 lb $H_2SiF_6$/month) × (1 gal/10.5 lb) = 662.26 gal/month

**Table 6-1. Comparison of Cost and Properties of Fluoride Compounds**

| | *1993 Estimates* | | |
|---|---|---|---|
| *Name* | *Sodium Fluoride* | *Sodium Silicofluoride* | *Hydrofluosilicic Acid* |
| Formula | NaF | $Na_2SiF_6$ | $Na_2SiF_6$ |
| Purity range | 95–98% | 98–99% | 23–30% |
| Formula weight | 42 | 188 | 144 |
| % available F- ion | 45% | 61% | 79% |
| Cost per Gal (1993) | Not applicable | Not applicable | $0.975 @ 10.5 lb/Gal |
| Cost per lb (1993) | $0.60 | $0.27 | Not applicable |
| Cost per lb. F- ion | $ 1.33 | $0.44 | $0.51 @ 25% |
| Cost/MG for 1 mg/L Dosage | $11.09 | $3.72 | $4.25 |

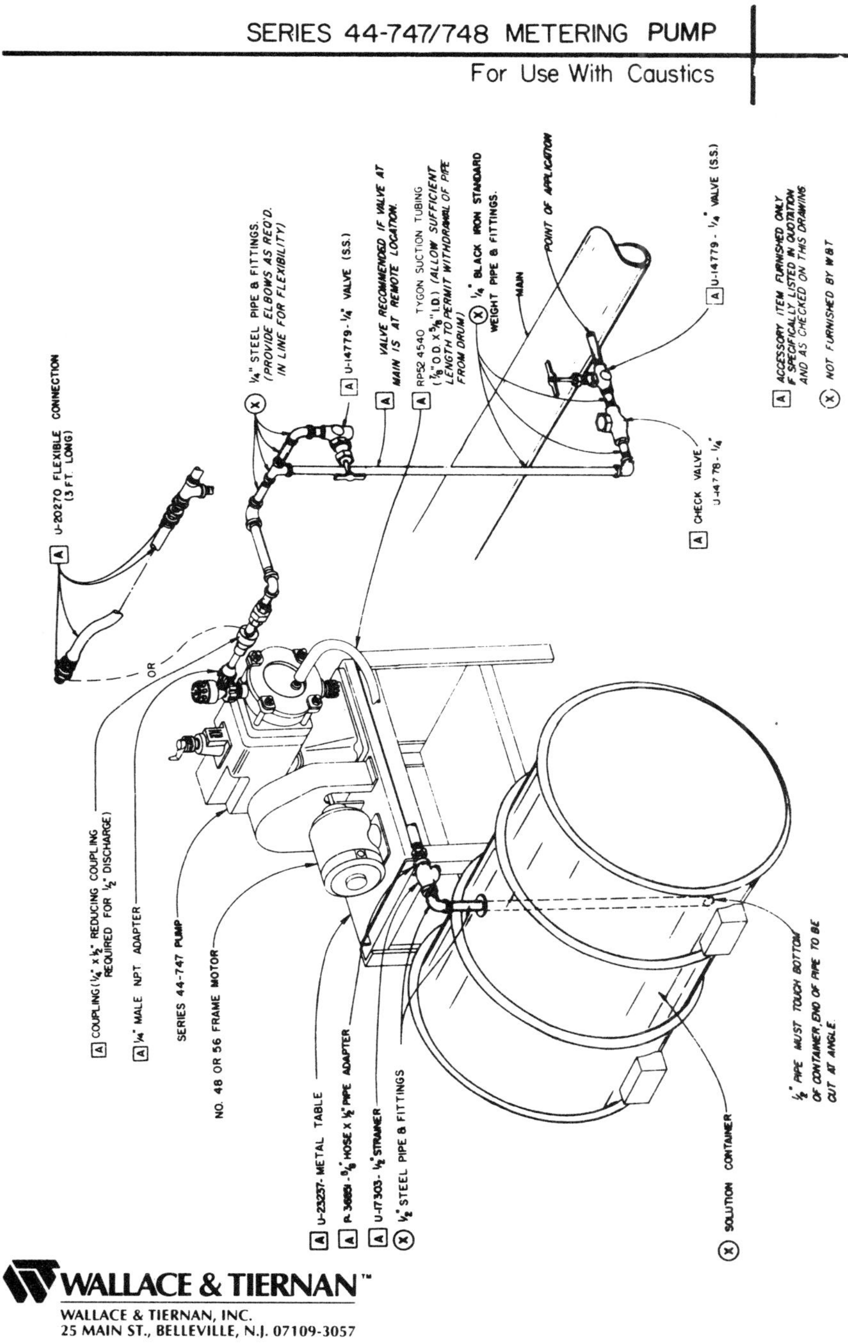
SERIES 44-747/748 METERING PUMP
For Use With Caustics
A U-20270 FLEXIBLE CONNECTION (3 FT. LONG)
X 1/4" STEEL PIPE & FITTINGS. (PROVIDE ELBOWS AS REQ'D. IN LINE FOR FLEXIBILITY)
A U-14779 - 1/4" VALVE (S.S.)
A VALVE RECOMMENDED IF VALVE AT MAIN IS AT REMOTE LOCATION.
A RP52 4540 TYGON SUCTION TUBING (7/8" O.D. X 5/8" I.D.) (ALLOW SUFFICIENT LENGTH TO PERMIT WITHDRAWAL OF PIPE FROM DRUM)
X 1/4" BLACK IRON STANDARD WEIGHT PIPE & FITTINGS.
MAIN
POINT OF APPLICATION
A U-14779 - 1/4" VALVE (S.S.)
A ACCESSORY ITEM FURNISHED ONLY IF SPECIFICALLY LISTED IN QUOTATION AND AS CHECKED ON THIS DRAWING
X NOT FURNISHED BY W&T
A CHECK VALVE U-14778 - 1/4"
OR
A COUPLING (1/4" x 1/2" REDUCING COUPLING REQUIRED FOR 1/2" DISCHARGE)
A 1/4" MALE NPT ADAPTER
SERIES 44-747 PUMP
NO. 48 OR 56 FRAME MOTOR
A U-23257- METAL TABLE
A P-36851 - 5/8" HOSE X 1/2" PIPE ADAPTER
A U-17303 - 1/2" STRAINER
X 1/2" STEEL PIPE & FITTINGS
1/2" PIPE MUST TOUCH BOTTOM OF CONTAINER, END OF PIPE TO BE CUT AT ANGLE.
X SOLUTION CONTAINER
WALLACE & TIERNAN™
WALLACE & TIERNAN, INC.
25 MAIN ST., BELLEVILLE, N.J. 07109-3057

FIGURE 6-2

Chapter VII

# Tastes and Odors

Obnoxious tastes and odors may be found in supplies which are susceptible to the growth of certain blue-green algae and/or actinomycetes. The metabolic activities of these microorganisms can produce chemicals called geosmin and methylisoborneol (MIB) which are known to be detectable by humans in very small amounts (nanograms/Liter). When odors are due to these two agents, they often come on unexpectedly and can be most difficult to remove.

Tastes and odors in water may be caused by:

1. Dissolved organic matter.
2. Dissolved organic gases.
3. Hydrogen sulfide.
4. Mineral substances, such as sodium chloride, sodium sulfate, magnesium sulfate, and iron.
5. Earthy constituents.
6. Algae and microorganisms.
7. Phenols and other oily or tarry waste products.
8. Chlorine products, especially chlorinated hydrocarbons.

Methods used for reducing or removing tastes and odors are:

1. Aeration.
2. Coagulation (using excess coagulants and/or coagulant aids).
3. Lime softening.
4. Break-point chlorination.
5. Chlorine dioxide.
6. Ammonia.
7. Permanganate.
8. Adsorption by Activated Carbon.
9. Ozone.
10. Various combinations of the above.
11. Copper Sulfate

Water containing mineral matter in sufficient quantity to produce an unpleasant taste seldom is fit for domestic use. Almost all other tastes and odors can be removed by careful use of the proper kind and amount of treatment.

Aeration is effective for removing tastes due to dissolved gases but helps little in oxidizing organic matter. However, iron is oxidized and made insoluble so that it can be removed. Coagulation, using larger quantities of coagulant than required for clarification only, sometimes is effective in removing or reducing tastes due to earthy constituents.

When sufficient lime is used to soften the water, organic matter is more effectively removed than under ordinary conditions of coagulation, thus assisting in the elimination of tastes and odors. Lime softening can produce a bacteriological safe water without the use of chlorine. Lime softening also has been used to remove tastes and odors due to sulfides.

Because of the excellent bacterial quality of lime softened water, it is not only possible but practical to produce a finished water free of any noticeable chlorinous taste. Such chlorine dosage that is needed for additional safety is minimal, and is sometimes combined with ammonia to form chloramine.

Copper Sulfate use for algae control in surface water is a common time-tested practice. Dosages vary according to the types of reservoirs being treated, but caution must be used as many fish are susceptible to the chemical even at low dosages. Application of the chemical is accomplished by dragging porous bags of it by boat across the water, or by spraying a dissolved solution of chemical on the surface of the supply.

Potassium Permanganate is used by many suppliers as a substitute for Copper Sulfate in the battle against taste and odors. The Lake County, Ohio, Aquarius Treatment Plant feeds $KMnO_4$ at the mouth of the lake intake at a dosage of 0.5 mg/L. This allows for a 35 minute detention period before coagulation, during which time any algae found in the lake are inactivated. Care must be used against overfeed because of the color produced. Permanganate produces an insoluble end-product, $MnO_2$, which is removed in the filtration process. $KMnO_4$ will also destroy phenols, cyanide, and some color. It does not remove certain difficult to oxidize taste and odor compounds, and it does not totally destroy total trihalomethane (TTHM) precursors.

In 1987, only five plants in the U.S. were using Ozone primarily for taste and odor control. Ozone appears to be effective in removing or destroying odors originating from geosmin and methylisoborneol, but can be somewhat limited. Care must be taken with Ozone as it can, in some cases, actually increase the TTHM formation potential.

No fixed rule can be given for removing tastes and odors from a water supply. Each individual water is a problem unto itself. The best way to determine the proper treatment is by trial in the laboratory. Laboratory methods are given in Chapter XVIII.

PICTURE 7-1. AERATORS AT MARSTON LAKE, DENVER, COLORADO

One of the most extensively used procedures for taste and odors is the application of activated carbon, which is quite effective in many cases, particularly when algae and microorganisms are responsible. Handling and feeding the material is usually a disagreeable job, but recent developments in slurry storage and feeding have improved this to a great extent. Use of air slide trucks for hauling has made bulk delivery and storage of carbon in the dry form, and especially as a slurry, much more desirable since the unloading problems are practically eliminated.

While the carbon may be applied together with the coagulant, or to the settled water just prior to filtration, neither of these points may be too desirable. In the first point (with coagulants) the carbon is removed by the floc before it has sufficient time to adsorb the objectionable taste and odor material, and at the second point, it is removed by filtration, before it has adequate contact time. If the plant design permits two stages of sedimentation (in series) application of carbon at the influent to the second basin may give the best results.

If it is possible to do so, application of the carbon prior to any other chemical feed may be the most efficient. One plant has had good results by application of carbon at a low service pumping station; thus allowing about three hours contact time before addition of coagulant and the lime and soda.

Another factor to consider is that carbon will adsorb chlorine, so if the two materials are added at or near the same point, the effectiveness of both are very greatly reduced.

Break-point chlorination may be used to decompose or break down some taste- and odor-forming compounds, and is generally effective for phenols, but not in all cases. It must be used properly so that the dosage does not fall below that required for break-point. Also if the dosage is too high, a dechlorinating agent may be required.

Chlorine dioxide has been even more effective in some cases, due to its higher oxidizing power. It has been shown to be effective in removing phenolic tastes, especially where these tastes have been "set" by lime softening. The disadvantage lies in the costs of sodium chlorite and in the hazards involved.

Ammonia is used to combine with chlorine to reduce chlorinous tastes. This also has been found effective in treating well supplies, particularly where the distribution system becomes fouled with odors created by sulfur-reducing bacteria. Break-point chlorination can cause an excessive oxidation potential, thus intensifying the red-water troubles by loosening old scale and tubercles. Chloramines can effectively hold the bacterial growth in check without intensifying the iron difficulties.

## Chapter VIII

# Prevention of Red Water and Corrosion

As iron is slightly soluble in water, it probably is safe to say that all public water supplies would carry some of it in solution, and be unsatisfactory for domestic consumption, if the water were not prevented from coming in direct contact with the iron of the distribution pipes. The higher the acidity of the water, or the lower the pH value, the higher will be the amount of iron dissolved. Water with a pH of 9.0, for instance, dissolves only traces of iron, whereas water with a pH of 6.0 or 6.5 dissolves appreciable quantities. A pH around 7.5 to 8.0 has less effect on zinc solubility than higher pH values.

Red water troubles in a distribution system can be prevented by adjusting the chemical balance of the water. Water may be treated so that an impervious film will be deposited on iron surfaces with which it comes in contact. In this manner iron surfaces may be protected against the dissolving action of water. Water that is supersaturated with calcium carbonate will quickly deposit a coating of film on the inside of the pipe. Water that is almost saturated, and this applies to numerous natural supplies, also will deposit a protective coating, because in the process of corrosion carbon dioxide is taken from the water. The solubility of calcium carbonate in natural water depends on the carbon dioxide content; therefore, if some of the carbon dioxide is absorbed, the nearly saturated water becomes saturated, or perhaps supersaturated. When this point is reached, deposition of calcium carbonate starts and the corroding iron becomes coated with a protective film of calcium carbonate. Water that is undersaturated with calcium carbonate will dissolve calcium carbonate film already formed.

Whether or not water is in chemical balance to calcium carbonate can be easily and quickly determined. (See Calcium Carbonate Chemical Balance or Stability Test, Chapter XVIII.) Some metals, such as wrought iron, do not build up a calcium carbonate coating readily, even though the water shows a fairly high calcium carbonate excess by the stability test.

In using the chemical balance process to build up a protective coating, care must be taken not to continue the building up process too long because pipes will become clogged. This has happened in some municipalities having lime-soda ash water softening plants. The water produced was supersaturated which resulted in too much deposit in the pipe lines, meters, and heating coils, and necessitated finding corrective measure. A practical corrective treatment process has been developed and the deposition now is controllable by the process of recarbonation; that is, by adding carbon dioxide gas to the water. See Chapter XIII. Protective films built up by calcium carbonate will not last long if the water brought in contact with them is undersaturated, as this condition will dissolve the film.

This condition can be remedied by the addition of small quantities of sodium polyphosphate $(NaPO_3)_6$ (0.25 to 1.0mg/l) to the water. The water will then be stabilized, and is evidenced by the fact that alkalinity determinations made on samples taken from far distant points in the distribution system show approximately the same results as samples collected at the plant. This treatment permits a satisfactory protective scale to be formed in galvanized hot water tanks provided the temperature does not exceed 140° to 150°F.

If the analysis shows that the water is undersaturated, then it is necessary, if red water troubles are to be prevented, to build up the calcium carbonate content. This can be accomplished by adding lime to the water, or if the water already contains some calcium sulfate, soda ash can be added and the results will be satisfactory. Soda ash reacts with calcium sulfate to form calcium carbonate. It may be convenient to build up the calcium carbonate content by adding some hard water.

After the calcium carbonate and pH adjustments have been made, it is easy to determine whether or not the water is saturated, supersaturated, or undersaturated by making a calcium carbonate chemical balance test, or by applying the Langelier formula described in Chapter XVIII.

Corrosion tests are so easily and so quickly made that any operator should be able to determine without difficulty whether or not the water produced at his plant is in chemical balance. Further discussion of this subject will be found in Chapter XIII.

Corrosion is sometimes prevented, or at least retarded, by other protective coatings. Often, water plants that do not employ the lime soda process will resort to the addition of silicates or orthophosphates to the supply. Water chemistry profiles are necessary for making informed decisions about the use of such chemicals. For example, the use of orthophosphates can dramatically reduce lead solvency in a system given the proper dissolved inorganic carbon levels (DIC), the correct pH, and the proper temperature and dosage; but in some cases can actually increase lead solvency.

The advent of the Lead and Copper Rule of the SDWA (1992) has forced the water supply industry to reconsider its definition of a noncorrosive water. Now systems must sample water from a certain percentage of their customers' taps. Ninety percent of these samples must not exceed the Action Level (AL) of 15 micrograms per liter lead and 1300 micrograms per liter copper. If they do, the water supply is considered corrosive regardless of its Langelier Index, pH, hardness, etc. In other words, it may be considered to be in chemical balance by traditional means; nonetheless, it is corrosive if lead and copper dissolve in it in quantities which exceed the AL's.

Operators of water systems should have little difficulty meeting these AL's if the following steps are performed:

1. Lime soda plants must treat water to a slightly positive pH by some acceptable index (Langelier, Ryznar, etc.).
2. Conventional plants should add some inhibitor (Zinc orthophosphate, e.g.) if pH is kept below 7.6.
3. All system water mains, or at least the dead ends, should be flushed twice per year at velocities exceeding 2.5 fps, especially if inhibitors are used. Do not sample for lead and copper until at least thirty days after system flushing.
4. Well water supplies which are softened by the zeolite process or are aerated will probably need some treatment as mentioned in 1 or 2 above.
5. A system of coupon or corrator testing should be employed, and a mils loss per year rate of no lower than one (0.001 inch) and no higher than three (0.003 inch) should be the target corrosion rate (for most systems). Plant operators are cautioned against achieving a very low (<1 mil/yr) corrosion rate. This rate has been suggested as a deterrent to inactivation of system biofilms. When too much deposition occurs, chlorine does not readily transport to the pipe wall where it can be utilized.
6. Assimilable Organic Carbon (AOC) should not exceed 50 micrograms per liter throughout the distribution system as it will encourage and support the growth of biofilms on the pipe walls. Biofilms act as galvanic cells which promote corrosion. AOC can be kept low by consistently producing a low turbidity water.
7. A public education program, aimed at teaching the customers how to minimize lead and copper levels in their home water supplies, should be instituted. This is the information age, and the public should be involved in affairs of the water department in their community. An uninformed public can fall prey to scare tactics of an unscrupulous water conditioner salesperson.

The reader is referred to Lead Control Strategies, a publication of the American Water Works Research Foundation.

Chapter IX

# WATER SOFTENING

## Lime and Soda Ash Process

### *What is Hard Water?*

The consumer judges water to be hard if much soap is required to produce a lather, and soft if it lathers freely. The hardness of most water supplies is due to four compounds which are held in solution.

These compounds are:

1. Calcium bicarbonate (by solution of limestone, $CaCO_3$, in waters containing carbon dioxide, $CO_2$).
2. Magnesium bicarbonate (solution of magnesite, $MgCO_3$, in waters containing carbon dioxide, $CO_2$).
3. Calcium sulfate (in the form, $CaSO_4 \cdot 2H_2O$, is known as gypsum).
4. Magnesium sulfate (in the form, $MgSO_4 \cdot 7H_20$, is known as epsom salts).

In addition to the sulfate forms, chlorides and nitrates of calcium and magnesium are sometimes present.

As long as any of these compounds or a combination of them are present in a water supply it is hard.

Calcium and magnesium bicarbonates formerly were designated as temporary hardness because they could be partially precipitated by boiling the water to drive off carbon dioxide, but now are termed **carbonate hardness**.

Calcium and magnesium sulfates, chlorides, and nitrates formerly were designated as permanent hardness, but now are called **noncarbonate hardness**.

Iron and manganese, and other divalent cationic ions also cause hardness, but are generally present in such small quantities that it is customary not to consider them in connection with hardness. However, the presence of iron and manganese is very objectionable. They stain materials with which they come in contact. The stain is noticed particularly in washbowls, bathtubs, sinks, and lavatories. They are especially objectionable if the water is used for laundry purposes. Iron and manganese can be removed separately using oxidation treatment processes

(See Chapter XII), but the lime water softening process also removes iron and manganese as well as hardness salts.

## Where Hard Water Is Found

Most hard water supplies are associated with systems utilizing groundwater as their raw water source, although it is not uncommon to encounter surface water supplies that are hard. Much of the hardness in water comes from the water contacting soil and rock formations. Soil water becomes high in carbon dioxide content, from the release by bacterial action of carbon dioxide that reaches equilibrium with the carbonic acid content of the water. Under these low pH conditions limestone, dolomitic limestone and other basic materials are dissolved by the soil water.

More than two thirds of the states in the U.S. have hard water, with seven of them—Indiana, Iowa, South Dakota, Nebraska, Arizona, New Mexico, and Utah—having extremely hard water. The soft water states include those in New England, New York, the Southeast (except Florida), Arkansas, Oregon and Washington.

The degree of hardness by states is shown in Fig. 9-1, which gives the weighted average hardness of water furnished by public water systems in the larger U. S. cities. The figures are weighted according to population, and represent averages; it is not to be construed that every supply in the heavily shaded areas is necessarily hard.

## How Hard Water Is Softened

In general terms, hard water is made soft by: 1) a treatment which changes calcium and magnesium compounds from a soluble to an insoluble form, and then by removing the insoluble compounds by sedimentation and filtration; or 2) by an exchange process which consists of replacing the divalent calcium and magnesium ions with monovalent sodium ions. The first process is known as the chemical, or lime-soda process, and the second, the zeolite (ion exchange) process. A modified reverse osmosis membrane (i.e., the nanofiltration membrane) has been used in recent years to soften water, but this process has not yet gained wide acceptance.

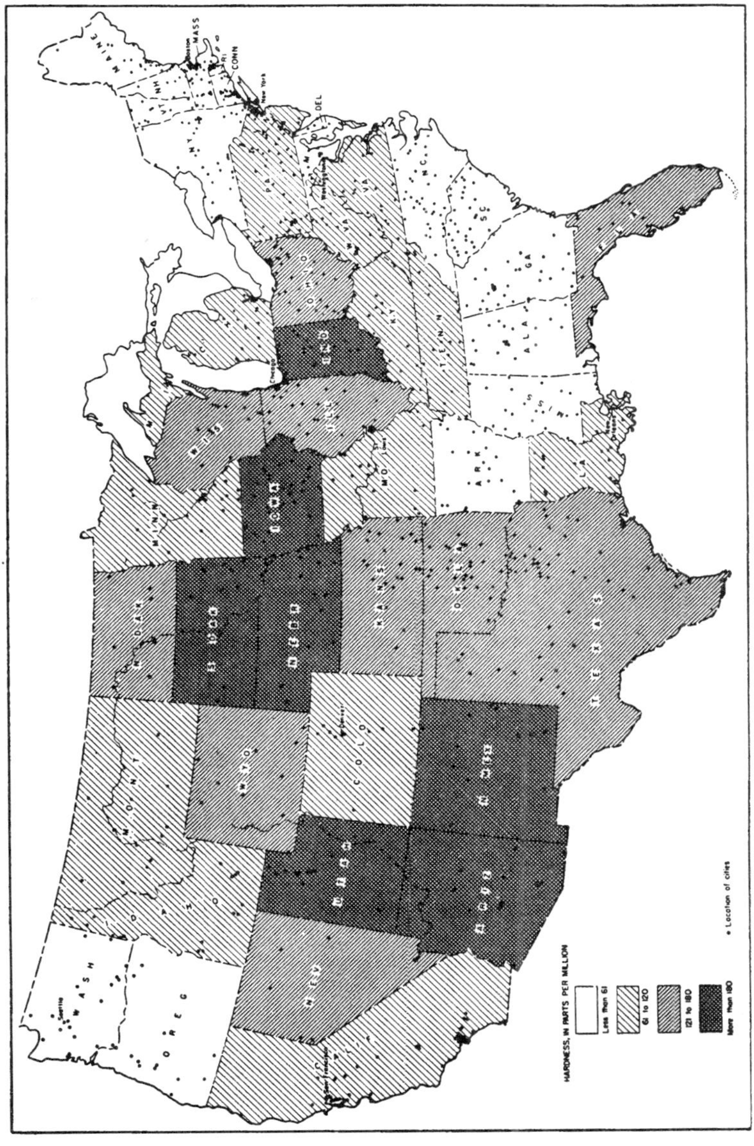


FIGURE 9-1. WEIGHTED AVERAGE HARDNESS BY STATES, OF FINISHED WATER FROM PUBLIC SUPPLIES FOR 1,315 OF THE LARGER CITIES OF THE UNITED STATES, 1952. (U.S. GEOLOGICAL SURVEY, WATER—SUPPLY PAPER 1300, PLATE 1)

## History of the Lime-Soda Ash or Chemical Process

The art of water softening with lime dates back to the year 1766 when Cavendish discovered that the addition of lime to natural water caused a deposition of the carbonates of calcium and magnesium. The proposal to use lime for softening hard water on a plant scale was advanced by Thomas Henry in England about 1800, but it was not until 1841 that water softening on a large scale was developed by Dr. Thomas Clark of Aberdeen, Scotland. He patented the process of adding lime to water to precipitate earthy carbonates held in solution by carbonic acid. About 1856, Dr. Porter, of London, suggested the use of soda ash for removing non- carbonate hardness. The lime-soda ash process for years was referred to as the Clark-Porter process. The Clark-Porter process developed in response to the need for soft water by the increasing number of industrial plants employing steam for power. There is not much recorded information on the number of softening plants built from 1841 to 1893 but the records show that from 1893 to 1898, upwards of fifty softening plants were built in England and abroad, including both public supplies and industrial plants with ranges in capacity of 12,000 to 1,000,000 gallons per day.

PICTURE 9-1. LIME-SODA SOFTENING PLANT, COCOA, FLA, SERVING CAPE KENNEDY SPACE CENTER

In this country, a number of plants for softening water for boiler feed and laundry purposes were built prior to 1901, at which time the first municipal water softening plant on this continent was built at Winnipeg, Canada. In 1903, the first municipal plant in the U.S. was built at Oberlin, Ohio.

In 1908, water softening on a larger scale than had previously been tried was undertaken at Columbus, Ohio, when a plant having a capacity of thirty million gallons per day was constructed. This plant, which was enlarged to a fifty-four million gallon capacity in 1923, had the distinction, for a number of years, of being the largest water softening plant in the world using both lime and soda ash.

The Chain of Rocks plant, St. Louis, MO, having a capacity of approximately 100 million gallons daily, has effected partial softening since its construction in 1903. The addition of lime and sulfate of iron to the turbid water of the Mississippi was originally intended for clarification purposes, softening being a secondary consideration. In 1915, filters were added to the Chain of Rocks plant. This increased the capacity to 160 million gallons daily.

For years the advantages of water softening were offset to a certain extent by disadvantages. Difficulties arose in operating sand filters on account of incrustation of the sand with calcium carbonate. Water mains, service pipes, and meters became clogged with heavy deposits. When water contained an excess of causticity it was unpalatable. The recarbonation process developed by Archbutt and Deeley in the latter part of the last century was employed at several plants in England for the purpose of overcoming all the above difficulties, but its first use in the U.S. on a successful plant scale was in 1921 at Defiance, Ohio. Since that time practically all water softening plants have been equipped with recarbonation devices.

The chemistry of water softening now is well understood, and accurate predictions of the results to be obtained can be made. Difficulties encountered on account of incrustation and excess causticity have been eliminated by recarbonation, flood waters can now be satisfactorily clarified and softened, and finished water of almost any degree of hardness can be produced.

Practically all laundries, power houses, ice plants, canning factories, railroads, and other industrial plants have installed softening plants where a soft water is required to improve or lower the cost of their product.

## Explanation of Lime Treatment

### *Water Not "Doped" with Chemicals*

"The water is doped with chemicals" is an expression which was often heard. Many laymen believed that lime made water hard, but this is not true unless too much or too little lime is added in the treatment process. A brief explanation of what happens when lime is added to hard water answers the question, "Why add lime, which contains calcium, to water to soften it?" so often asked by water works chemists, and also verifies why the addition of lime does not "dope" the water.

Lime, when added to water in the proper proportions, is itself completely precipitated and, in addition, precipitates the soluble calcium and magnesium salts already present in the water being treated. These precipitates may be removed easily by sedimentation or filtration. Details are explained in Chapter X. For each pound of lime added to hard water approximately three and one-half pounds of solids are precipitated. It is a process whereby something is removed from the water rather than being added.

## Does Softening Water Affect the Taste?

The removal of hardness salts alters the taste of water very little and has no bad effect on the health of the public. People quickly become

PICTURE 9-2. MIAMI FLORIDA WATER TREATMENT PLANT

accustomed to the difference. The percentage of salts in hard water is relatively small compared to the salts in other foods and drinks. Milk contains approximately twenty-five times as much mineral salt as moderately hard water, and the normal diet adjusts automatically to any loss suffered by the use of soft water. People in cities and states having access to a naturally soft water supply certainly are no less healthy than residents of hard water regions.

The following experiences demonstrate that softening of well water does not spoil its taste: At a banquet where 135 men and women were in attendance, two glasses of water were placed at each plate. One glass contained unsoftened hard well water, and the other, water taken from the same supply but softened to about 70 mg/L as $CaCO_3$. They were designated by placing a rubber band around the glass containing the hard water. One hundred and one said they could taste no difference between the hard and softened waters. Nineteen out of the remaining thirty four said the hard water was soft and fifteen guessed correctly. In other words, more of them guessed wrong than right.

At another city a blindfold test was made. Members of the city council were asked to taste, while blindfolded, water direct from the tap, and water from the same source which had been softened with lime. The fact they were convinced the softening process would not spoil the taste of the water was evidenced when they then passed the necessary legislation for a softening plant.

PICTURE 9-3. LIME-SODA WATER TREATMENT PLANT

At a banquet in another city more than 100 were in attendance. A glass of softened well water was placed in front of each fifth plate at the table. Upon questioning, no one of this gathering had noticed anything different about the water.

## Economic Aspects of Water Softening

### *Why Soften Water?*

Hard water should be softened because:

1. Hard water consumes soaps and detergents.
2. Hard water has an adverse effect on clothing and other articles being cleansed.
3. Hard water shortens the life of pipes and fixtures, heating systems, and boiler shells and tubes.
4. Hard water is unsuitable for many industries.

Studies have shown that it is more economical to soften the entire supply at the municipal treatment plant (even though only a portion of the water is used in the home where softening is needed), than to soften in the individual homes or even in the industrial plants. In the home, softening may be accomplished by: (1) the use of soaps and detergents in larger quantities, (2) a home-owned ion exchange (zeolite) softener, or (3) a serviced softener.

It has been recognized for years that use of soap is the most costly way to soften water, but the development of synthetic detergents (or syndets), and the publicity that has accompanied their use, built up the erroneous idea that the cost balance could be reversed. Most syndets now marketed contain a large proportion of builders such as phosphates, silicates, or carbonates which in reality are softeners, so the active detergent can be more effective.

A very comprehensive report* shows that syndets, while a decided improvement over soaps, do not perform as satisfactorily in hard water as in soft, and that softening still is needed. In this report these statements by chemists and technicians are given:

(1) The detergents do not perform quite as well in hard water—this situation becomes worse as the hardness increases.

(2) Hardness ions interfere and cause some of the dirt to be redeposited in the fabric. The real answer is soft water to begin with.

---

*Aultman, W.W., Journal, AWWA, Vol. 50, p. 1353, Oct. 1958.

(3) A greater amount of synthetic detergent is needed in hard water.

(4) In soft water a 0.15% concentration (of syndet) would be needed for cleaning while in 21 grain hardness, probably a 0.3% concentration would be required.

(5) You (may) assume that softening fairly hard water would save the average family 1/3 to 1/2 carton of detergent (20-22 oz. size) per week.

In the same report Aultman shows the results of a joint study made in 1955 by Dow Chemical Company and Woman's Home Companion under supervision of Purdue University. In this case a family of five used: (a) hard water and syndets, (b) soft water and syndets, (c) hard water and soap, and (d) soft water and soap. Each test ran for six weeks, with the cost data being projected to an annual basis.

| | |
|---|---|
| Annual costs of syndets with hard water | \$63.90 |
| Annual costs of syndets with soft water | \$30.24 |
| Annual saving of syndets, using soft water | \$33.66 |
| Annual costs of soaps, with hard water | \$84.13 |
| Annual costs of soaps, with soft water | \$32.91 |
| Annual saving of soap, using soft water | \$51.22 |

It is estimated that these costs, also the savings, would be significantly more with the present prices of soaps and syndets.

In the above observations, the hard water contained 380 mg/L hardness as $CaCO_3$; the soft water had zero hardness. Since municipal plants usually remove all but 85 to 100 mg/L hardness, calculation was made on the savings to be expected if the hardness of the water was reduced from 380 to 100 mg/L. These costs showed that the savings to the individual consumer for the purchase of less cleansing agents far outweighed the costs to the consumers of softening the water at a municipal water plant.

Water treatment authorities are becoming more and more concerned about the increasing amount of syndets in their water supplies. While the problem has not yet become critical, it is possible that if the increase in syndets in the water continues at the present rate, the coagulation processes may be adversely affected. There is no economical means, at the present time, to effectively remove these substances; the best answer seems to be in limiting their use in the first place. Softened

PICTURE 9-4. SLUDGE BLANKET PROCESS, MINNEAPOLIS WATER PLANT.

water, which reduces the required amount of syndet, is one means of keeping the available water supplies in better condition.

## Softeners in the Home

Where the municipal supply is hard, the home owner often resorts to treatment of his own supply, either by: 1) purchasing his own softening unit, or 2) securing serviced softening in which an outside agent installs a regenerated softening unit and then replaces it with a freshly regenerated unit at specified intervals.

Home-owned units are generally less expensive to maintain than the serviced softening units, with the only on-going costs being the salt for regeneration. However, the amortization costs of the home-owned softening unit, depending on the size of the family and the hardness of the water, must be considered as part of the cost.

The costs for serviced softening units vary widely from place to place. However, there is no original cost to purchase the unit that must be amortized in this case. Beside the expense of softening by this means, the consumer must use the water at a fairly uniform rate, or else run the risk of either paying for unused capacity at low consumption or having unsoftened water during periods of high consumption.

Based on recent research, a new type of home softener is becoming available. It is a membrane process, similar to reverse osmosis units, that selectively removes calcium and magnesium from hard waters. This membrane unit eliminates the need to add salt for regeneration, but creates a brine solution that must be discharged from the unit as a waste stream.

## Cisterns

Cisterns are a very unsatisfactory substitute for a municipal softened supply. Such water is not always soft, seldom adequate, and usually dirty, turbid, and highly colored. In addition, this water usually has an unpleasant taste and often is polluted with harmful bacteria. Cisterns are also an added expense in construction, maintenance and operation.

## Other Advantages of Softening

The softening process has been included in many new municipal water treatment plants, particularly where the hardness of the water exceeds 150 mg/L. This process has been adopted primarily to obtain the added comforts and benefits from soft water, but in some instances the deciding factor has been the increased efficiency of the filtration plant resulting from softening. Better coagulation is effected because the bulk of the precipitate formed may be from ten to twenty times as great as could be obtained by ordinary coagulation processes.

The efficiency of this large volume of precipitate as a coagulant is greatly increased if the water treated contains magnesium compounds. The lime reacts with these compounds to form magnesium hydroxide, a bulky gelatinous precipitate which coagulates the suspended clay particles or mud as readily as the hydroxide precipitates formed by sulfates of aluminum or iron. The precipitation of magnesium also reduces the fluoride and silica content of the softened water. If the quantity of magnesium salts in the water being softened is low, then a small quantity of coagulant may be added with the softening chemicals to make the precipitate more gelatinous and more efficient.

A properly lime softened water should have a bright greenish appearance and a turbidity less than 1 to 2 NTU when it leaves the settling basins. Such a water, in addition to being low in turbidity, should be low in color and organic matter and practically free from intestinal and pathogenic bacteria. The load on the filters is greatly lessened and filter costs are reduced because filtration rates can be increased. The load applied to the filters should be so light that filter runs can be lengthened to at least 48 hours between washings. At one lime softening plant, filters on special occasions have been run continuously for periods of more than twenty-one days between washings.

Coagulation, filtration, and chlorination nearly always produce a water that is clear and safe for drinking purposes. Sometimes the amount of chlorine required to make the water bacterially safe is so large that the treated water is unpalatable because of disagreeable tastes and odors. If water contains a slight amount of phenol, or large amounts of organic matter, the addition of chlorine intensifies any tastes or odors present in the water. Chlorine may be omitted entirely during flood periods when the organic matter is high and the water tends to have an unpleasant taste or odor, if at these times an excess lime treatment is used to maintain 20 to 50 mg/L caustic alkalinity in the settled water for a 5-hour period. In addition to producing better coagulation, more organic matter and color are removed, and iron and manganese, if present, are eliminated, the treated water is noncorrosive, and intestinal and pathogenic bacteria are killed. Many small municipalities have wells that are too near sources of contamination and thus are potentially dangerous. If the water from such wells is hard it is good practice to soften it with lime. A perfectly safe water can be obtained. For small water plants, where the pumping is apt to be intermittent and only limited supervision is available, it has been shown that lime treatment is the safest treatment that can be adopted.

Still other advantages include:

1. A 25% saving is realized on fuel for heating soft water by the elimination of deposits of scale which retard heat flow.
2. An 18% saving is obtained on repairing, cleaning, and replacing plumbing that is caked.
3. Fabrics show 25% less wear and tear in soft water.
4. Soft water saves 16% wear and tear on utensils.
5. Food cooked in soft water retains its natural color and appearance, and its digestive properties. In making tea and coffee, 50% less leaves and grounds are used.
6. Soft water provides for better skin care, and eliminates the need for expensive bubble bath preparations. There is no bathtub ring. Shaving blades last 20% longer.

Chapter X

# CHEMISTRY OF LIME-SODA ASH PROCESS OF WATER SOFTENING

Reactions that take place when lime is added to a hard water containing both calcium and magnesium are:

$$Ca(HCO_3)_2 + Ca(OH)_2 \rightarrow 2CaCO_3 + 2H_2O \quad (1)$$
$$Mg(HCO_3)_2 + Ca(OH)_2 \rightarrow MgCO_3 + CaCO_3 + 2H_2O \quad (2)$$
$$MgCO_3 + Ca(OH)_2 \rightarrow Mg(OH)_2 + CaCO_3 \quad (3)$$
$$MgSO_4 + Ca(OH)_2 \rightarrow Mg(OH)_2 + CaSO_4 \quad (4)$$
$$MgCl_2 + Ca(OH)_2 \rightarrow Mg(OH)_2 + CaCl_2 \quad (5)$$
$$Mg(NO_3)_2 + Ca(OH)_2 \rightarrow Mg(OH)_2 + Ca(NO_3)_2 \quad (6)$$

Reactions involving the use of soda ash are:

$$CaSO_4 + Na_2CO_3 \rightarrow CaCO_3 + Na_2SO_4 \quad (7)$$
$$CaCl_2 + Na_2CO_3 \rightarrow CaCO_3 + 2NaCl \quad (8)$$
$$Ca(NO_3)_2 + Na_2CO_3 \rightarrow CaCO_3 + 2NaNO_3 \quad (9)$$

Reactions involving the use of caustic soda are:

$$Ca(HCO_3)_2 + 2NaOH \rightarrow CaCO_3 + Na_2CO_3 + H_2O$$
$$CaSO_4 + 2NaOH \rightarrow Ca(OH)_2 + Na_2SO_4$$
$$CaCl_2 + 2NaOH \rightarrow Ca(OH)_2 + 2NaCl$$
$$Ca(NO_3)_2 + 2NaOH \rightarrow Ca(OH)_2 + 2NaNO_3$$
$$Mg(HCO_3)_2 + 4NaOH \rightarrow Mg(OH)_2 + 2Na_2CO_3 + 2H_2O$$
$$MgCO_3 + 2NaOH \rightarrow Mg(OH)_2 + Na_2CO_3$$
$$MgSO_4 + 2NaOH \rightarrow Mg(OH)_2 + Na_2SO_4$$
$$MgCl_2 + 2NaOH \rightarrow Mg(OH)_2 + 2NaCl$$
$$Mg(NO_3)_2 + 2NaOH \rightarrow Mg(OH)_2 + 2NaNO_3$$

The major objective of adding softening chemicals (e.g., lime, soda ash or caustic soda) to a hard water is to convert the soluble calcium and magnesium compounds to the forms that are precipitates (i.e., calcium carbonate, $CaCO_3$, and magnesium hydroxide, $Mg(OH)_2$). With this goal in mind, note that $MgCO_3$ is formed in reaction 2, and additional lime is added, reaction 3, to convert this $MgCO_3$ from reaction 2 to $Mg(OH)_2$. In reactions 4, 5 and 6; $Mg(OH)_2$ is formed, but the calcium from the lime forms soluble compounds (i.e., $CaSO_4$, $CaCl_2$ and

$Ca(NO_3)_2$). Therefore, soda ash must be added (reactions 7, 8 and 9) to convert these soluble compounds to $CaCO_3$.

The reactions involving the use of caustic soda, NaOH, are different than those with lime and soda ash. Note that when NaOH is added to react with calcium and magnesium bicarbonate, or magnesium carbonate; soda ash is formed.

The analytical data necessary to approximate the quantities of lime and soda ash required to soften a hard water containing soluble calcium and magnesium salts are:

1. Aqueous $CO_2$.
2. Alkalinity.
3. Total calcium and magnesium (total hardness).
4. Non-carbonate hardness (that not associated with alkalinity).

A rough estimate of the lime and soda ash requirements to soften a water completely can be determined by using the following rules-of-thumb:

1. For aqueous $CO_2$; lime must be added.
2. For calcium, carbonate hardness; lime must be added.
3. For magnesium, carbonate hardness; twice as much lime must be added.
4. For calcium, non-carbonate hardness; soda ash must be added.
5. For magnesium, non-carbonate hardness; lime and soda ash must be added.

The balanced chemical reactions can be used to approximate the lime and soda ash required to soften a water completely (i.e., to a minimum achievable total hardness of 40 to 60 mg/L, as $CaCO_3$). If a higher total hardness is desired in the finished water, less lime or soda ash is added than is required to soften the water completely.

A problem of using balanced chemical reactions to approximate lime and soda ash requirements for softening is that the molecular weights (g/mole) of the soluble calcium and magnesium salts and the lime and soda ash must be used. Therefore, the number of moles of each compound must be kept track of. A simpler concept is to use equivalent weights (g/eq), realizing that in softening reactions one equivalent of lime or soda ash reacts with one equivalent of the soluble calcium or magnesium salt, forming one equivalent of each product on the right-hand side of the chemical equation. Thus, the need to use a perfectly balanced chemical equation, and to keep track of the number of moles of each reactant or product is eliminated.

Three examples for approximating lime and soda ash requirements of typical plant operations are presented later in this Chapter, and an explanation of equivalent weights and the use of me/L graphs is provided in Chapter XVIII.

Caustic soda may be used to remove both carbonate and non- carbonate hardness from water. However, when caustic soda is added to water containing soluble calcium and magnesium salts, it precipitates the magnesium as insoluble magnesium hydroxide and absorbs carbonic acid, resulting in the formation of an equivalent amount of sodium carbonate (soda ash). If there is enough non-carbonate hardness in the water being treated to react with the sodium carbonate formed, caustic soda may be used as a softening agent instead of lime and soda ash. As a general rule, caustic soda can replace all the soda ash but only a part of the lime. With the higher cost associated with caustic soda, there would be little if any savings involved. Laboratory tests under these conditions indicate that the use of caustic soda may leave more calcium carbonate in the settled water than soda ash treatment. Use of caustic soda also involves hazards in handling material. Current market prices of caustic soda indicate that it does not always compete cost-wise with lime and soda ash.

Regardless of whether lime, soda ash or caustic soda is used to chemically soften a hard water; it must be remembered that the chemical reactions used to approximate dosages do not account for the purity of the softening chemical being added. The better commercial grades of quicklime are approximately 90 to 94 percent CaO, and high calcium hydrated lime is approximately 68 to 73 percent CaO. The best grades of soda ash are approximately 99 to 100 percent $Na_2CO_3$, and caustic soda is typically 98 to 99 percent NaOH. Therefore, quantities of pure chemicals approximated from calculations using the reactions should be divided by the appropriate percent purity and multiplied by 100 to determine the quantity of commercial-grade chemical that must be added.

It is well to check the calculations for required lime and soda ash by actually adding the calculated quantities to small measured quantities of the water to be treated. The following procedure has been found convenient. Make up a standard lime solution using 17.1 grams of lime in one liter of boiled and cooled distilled water. If quicklime is used, slake it with a small quantity of water, just enough to make a boiling solution, then dilute to one liter. This solution is of such strength that 1 milliliter added to one liter of water is equivalent to adding 1 grain per gallon. Standard solutions of hydrated lime or soda ash may be made according to the same formula.

Place one liter of the water to be treated in a large beaker and add the calculated quantities of lime and soda ash necessary to soften it. Stir gently for 30 minutes, allow to settle until the liquid becomes fairly clear (usually requiring 10 or 15 minutes), siphon off about half of the clear solution, and warm to a temperature of about 120°F, or 49°C. Filter through filter paper and then analyze to determine whether or not proper amounts of chemicals have been added.

Methods for making these tests are found in Chapter XVIII.

## Applications of Analyses to Plant Operation

### *Determination of Proper Quantity of Lime*

Determine the methyl orange (total) and phenolphthalein alkalinity of the treated water. If two times the phenolphthalein alkalinity (P) is more than the methyl orange alkalinity, M (i.e., 2P is greater than M), the water contains caustic alkalinity or excess lime. If two times the phenolphthalein alkalinity is less than the methyl orange alkalinity the water is undertreated with lime. The following table shows the relation between alkalinity to phenolphthalein and methyl orange:

| *Sample No.* | *Suppose the Alkalinity to M.O. to be** | *And the Alkalinity to Phenolphthalein to be* | *The analysis would show* | | |
|---|---|---|---|---|---|
| | | | *Bicarbonates* | *Normal Carbonates* | *Excess Lime or Causticity* |
| 1 | 44 | 0 | 44 | 0 | 0 |
| 2 | 44 | 20 | 4 | 40 | 0 |
| 3 | 44 | 22 | 0 | 44 | 0 |
| 4 | 44 | 24 | 0 | 40 | 4 |

(Results expressed in mg/L)

*Methyl orange indicator measures all the alkalinity, bicarbonates, normal carbonates, and hydroxide or caustic alkalinity. Phenolphthalein indicator measures none of the bicarbonate, all of the hydroxide or caustic alkalinity, and half the normal carbonates. It measures only half the normal carbonates because each ml. of titrating acid added to a solution containing normal carbonates releases an equivalent amount of carbonic acid which affects phenolphthalein the same as a titrating acid; therefore, if twice the phenolphthalein alkalinity exactly equals the methyl orange alkalinity ( i.e., 2P=M), all the alkalinity is present as normal carbonate. If two times the phenolphthalein alkalinity is more than the methyl orange alkalinity (i.e., 2P is greater than M), the water contains caustic alkalinity or excess lime, and if it is less than half, the water contains bicarbonates.

Sample No. 1 contains no lime.
Sample No. 2 contains too small a quantity of lime.
Sample No. 3 contains the theoretical correct quantity of lime.
Sample No. 4 contains caustic alkalinity or excess lime.

In the operation of a softening plant it is customary to carry excess or caustic lime to the extent of 10 to 50 mg/L (as $CaCO_3$). Water so treated, after recarbonation and filtration, should have an alkalinity of approximately 35 mg/L (as $CaCO_3$). Each 17 mg/L of caustic alkalinity represents approximately 1 grain per gallon excess lime.

*Determination of Proper Quantity of Soda Ash*

Determine the non-carbonate hardness. The results of this test will show whether or not the desired quantity of non- carbonate hardness, as calculated, has been removed. The quantity of non-carbonate hardness remaining in the water usually is regulated according to the adopted policy as to what total hardness is desired. If the total hardness is to be carried at 70 mg/L (as $CaCO_3$) it is necessary to remove all but 35 mg/L of the non-carbonate hardness. The 35 mg/L of non-carbonate hardness plus the probable reduction of the alkalinity to 35 mg/L would equal 70 mg/L of total hardness. If it is the policy to reduce the total hardness to only 80 mg/L, then all but 45 mg/L of the non-carbonate hardness should be reduced.

*Determination of Proper Quantity of Caustic Soda*

Caustic soda can be used in place of both lime and soda ash, but its reacting equivalent is in a fixed ratio (in approximate figures: 4 lbs. of caustic soda can replace about 3 lbs. of quicklime and 5 lbs. of soda ash). However, unless the water softening reactions require lime and soda ash in the ratio of about 3.5, then either lime or soda ash feed would be required in addition to the caustic soda.

If the computed lime to soda ash feed is less than 3 lbs. of lime to every 5 lbs. of soda ash, caustic soda for the lime equivalent, plus additional soda ash, will be needed. Usually the feed is more than 3 lbs. of lime for every 5 lbs. of soda ash; thus lime plus caustic soda feed would be indicated.

## Expression of Water Softening Results

The chemistry involved in the determination of alkalinity to methyl orange, alkalinity to phenolphthalein, and non- carbonate hardness, is extremely interesting because from these results the composition of water, expressed in the following terms, may be calculated:

Carbonate hardness  
Total hardness.  
Bicarbonates of calcium and magnesium.  
Normal carbonates of calcium and magnesium.  
Caustic alkalinity as calcium hydroxide.  
Caustic alkalinity as sodium hydroxide.

This calculation is of particular interest in water softening when excess lime and soda ash are used to produce water of low degree hardness.

Caustic alkalinity is present if two times the alkalinity to phenolphthalein exceeds the alkalinity to methyl orange. Examples 2, 3, 5, and 6. (See following tables.)

If the total hardness of an alkaline water is greater than the alkalinity, the difference between the two equals non- carbonate hardness. Examples, 1, 2, 7, and 8.

In case of low pH acid waters, all hardness is non-carbonate hardness.

If the total hardness is less than the alkalinity, the water contains negative non-carbonate hardness. In other words, the water contains free sodium carbonate or sodium hydroxide. The difference between the alkalinity and total hardness equals negative non-carbonate hardness. Examples 3, 4, 5, and 6.

If the caustic alkalinity is greater than the negative non- carbonate hardness, the difference between them equals $Ca(OH)_2$. Example 3. Ca and Mg normal carbonates, $Ca(OH)_2$, and NaOH are present.

If the caustic alkalinity is less than the negative non- carbonate hardness, all the caustic alkalinity is NaOH, and the difference between the non-carbonate hardness and the caustic alkalinity equals $Na_2CO_3$. Example 6. Normal carbonates, NaOH, and $Na_2CO_3$ are present. Example 4 with no caustic shows Ca and Mg normal carbonates and $Na_2CO_3$.

If the caustic alkalinity is equal to the negative non- carbonate hardness, example 5, Ca and Mg normal carbonates and NaOH are present.

There are eight compatible combinates, four resulting with a positive non-carbonate hardness and four with a negative non- carbonate hardness. Graphically this is shown: *

* Indebted to Mr. W. A. Kramer, Assistant Chemical Engineer, Chain of Rocks Filtration Plant, St. Louis, MO., for this table. The use of a me/L graph (Chapter XVIII) makes it easier to group positive and negative ions of the raw water into hypothetical compounds that must be reacted with lime or soda ash to form either calcium carbonate or magnesium hydroxide precipitates for removal in the settling and filtration processes.

| *Non-carbonate Hardness* | + | + | + | + | - | - | - | - |
|---|---|---|---|---|---|---|---|---|
| Compatible Constituents | Bicarbonates $HCO_3$ | Bicarbonates and Normal Carbonate $HCO_3$ - $CO_3$ | Normal Carbonates $CO_3$ | Normal Carbonate and Calcium Hydroxide $CO_3$ - $Ca(OH)_2$ | Normal Carbonate, Sodium and Calcium Hydroxides $CO_3$, NaOH - $Ca(OH)_2$ | Normal Carbonates , Sodium Hydroxide $CO_3$ - NaOH | Normal Carbonate, Sodium Hydroxide and Sodium Carbonate $CO_3$ - NaOH - $Na_2CO_3$ | Normal Carbonate and Sodium Carbonate $CO_3$ - $Na_2CO_3$ |
| Example | 7 | 1 | 8 | 2 | 3 | 5 | 6 | 4 |

Note: By normal carbonates, calcium and magnesium normal carbonates are meant.

For further explanation of the above, the following examples are given. Eight samples of water show the following results of analysis, in mg/L:

| | *No.1* | *No.2* | *No.3* | *No.4* | *No.5* | *No.6* | *No.7* | *No.8* |
|---|---|---|---|---|---|---|---|---|
| Alkalinity to Methyl Orange | 58 | 70 | 70 | 60 | 60 | 60 | 90 | 60 |
| Alakalinity to phenolphthalein | 20 | 40 | 50 | 30 | 40 | 40 | 0 | 30 |
| Non carbonate hardness | 30 | 25 | –25 | –20 | –20 | –25 | 30 | 20 |
| Total Hardness | 88 | 105 | 45 | 40 | 40 | 35 | 120 | 80 |

**From the results of analysis the following may be reported (all in terms of $CaCO_3$)**

| | *No.1* | *No.2* | *No.3* | *No.4* | *No.5* | *No.6* | *No.7* | *No.8* |
|---|---|---|---|---|---|---|---|---|
| Carbonate Hardness | 58 | 70 | 45 | 40 | 40 | 35 | 90 | 60 |
| Non carbonate Hardness | 30 | 35 | 0 | 0 | 0 | 0 | 30 | 20 |
| Total hardness | 88 | 105 | 45 | 40 | 40 | 35 | 120 | 80 |
| Caustic Alkalinity | 0 | 10 | 30 | 0 | 20 | 20 | 0 | 0 |
| Bicarbonates of Ca and Mg | 18 | 0 | 0 | 0 | 0 | 0 | 90 | 0 |
| Normal Carbonates of Ca and Mg | 40 | 60 | 40 | 40 | 40 | 35 | 0 | 60 |
| Norm Carbonate of Sodium $Na_2CO_3$ | 0 | 0 | 0 | 20 | 0 | 5 | 0 | 0 |
| Caustic Alkalinity as $Ca(OH)_2$ | 0 | 10 | 5 | 0 | 0 | 0 | 0 | 0 |
| Caustic Alkalinity as NaOH | 0 | 0 | 25 | 0 | 20 | 20 | 0 | 0 |

## Procedures for Estimating Chemical Requirements

The reactions that take place when softening chemicals (e.g., lime, soda ash and caustic soda) are added to a hard water, and the application of analyses to the operation of a softening plant are discussed in the earlier sections of this Chapter. The concept of using me/L graphs to compare the cationic (positive) and anionic (negative) ions present in water is presented in Chapter XVIII.

The following three examples combine information from these two Chapters; providing an approach for approximating the amount of lime, soda ash and carbon dioxide required in the softening of water. These three examples cover the typical types of softening plants used in various communities (see Chapter XI):

1. Single-stage softening of well water in which the magnesium concentration of the raw water is not excessive (i.e., 50 mg/L, as $CaCO_3$, or less).
2. Two-stage, excess lime softening of surface water for which removal of turbidity, color and/or taste & odor compounds is also required—also applicable to well water for which a low magnesium concentration in the softened water is desired.

3. Split treatment softening of well water in a two-stage process—used to produce a softened water having a reasonable magnesium concentration, but by reducing the quantities of lime and carbon dioxide associated with two-stage, excess lime softening.

Comparison of calcium and magnesium hardness compounds with chemical dose requirements on a me/L basis has a distinct advantage. One milliequivalent of calcium or magnesium hardness reacts with one milliequivalent of the softening chemical in each of the chemical reactions presented earlier in this Chapter, and produces one milliequivalent of calcium carbonate or magnesium hydroxide precipitate. Thus, the confusion of using a balanced chemical equation, and comparing hardness compounds and softening chemical requirements on a molar basis is eliminated.

For illustrative purposes, and to provide a comparison of chemical requirements and the softened water characteristics, the same raw water characteristics are used for the following three examples. The results of the water analyses performed on this raw water are:

Positive Ions, Concentration (mg/L)

Calcium (Ca) = 120 mg/L
Magnesium (Mg) = 24.4 mg/L
Sodium (Na) = 23 mg/L
Potassium (K) = 39.1 mg/L

Negative Ions, Concentration (mg/L)

Bicarbonate ($HCO_3$) = 335.5 mg/L
Sulfate ($SO_4$) = 120 mg/L
Chloride (Cl) = 71 mg/L

Note: The raw water was also determined to contain 22 mg/L of carbon dioxide ($CO_2$) in solution.

To utilize the me/L graph for performing calculations to approximate the softening chemical dose requirements, the ion concentrations in mg/L must be converted to concentrations in me/L. This conversion is accomplished by dividing the mg/L concentration by the equivalent weight (mg/me) of each ion according to the following equation:

Equivalent Weight (E.W.) = mg/me
Conc. mg/L × (1/E.W.) = Conc. mg/L × (me/mg) = Conc. me/L

For the raw water used in the following three examples, the conversions are:

Calcium (Ca) = 120 mgCa/L × (me Ca/20 mg Ca) = 6 me Ca/L
Magnesium (Mg) = 24.4/12.2 = 2 me Mg/L
Sodium (Na) = 23/23 = 1 me Na/L
Potassium (K) = 39.1/39.1 = 1 me K/L
Bicarbonate ($HCO_3$) = 335.5/61 = 5.5 me $HCO_3$/L
Sulfate ($SO_4$) = 120/48 = 2.5 me $SO_4$/L
Chloride (Cl) = 71/35.5 = 2 me Cl/L
Carbon dioxide ($CO_2$) = 22/22 = 1 me $CO_2$/L

### *1. Single-Stage Softening Example*

The flow schematic of a single-stage softening facility is:

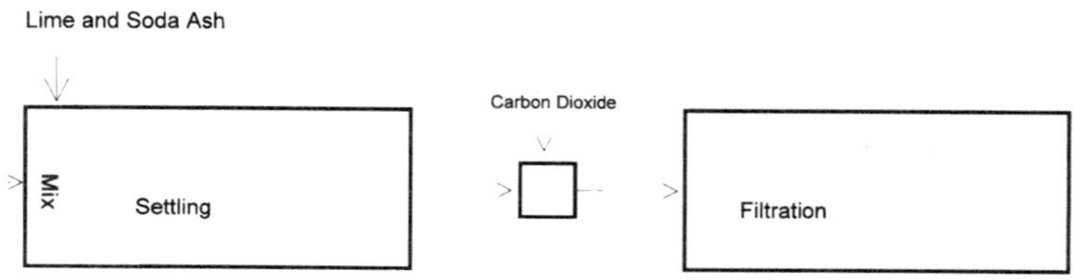


FIGURE 10-0. SINGLE-STAGE SOFTENING SCHEMATIC

Since the softening reactions take place in one reactor it cannot be operated at both the optimum conditions for calcium carbonate removal (pH value of 10.4) and for magnesium hydroxide removal (pH value of 11.2). Therefore, the lowest level to which the water can be expected to be softened is 30 mg/L (as $CaCO_3$) of calcium and 30 mg/L (as $CaCO_3$) for magnesium.

The **First Step** is to organize the positive and negative ions in a vertical me/L graph (Chapter XVIII), and to use the numbered chemical equations represented earlier in this Chapter to approximate chemical quantities required to soften the water.

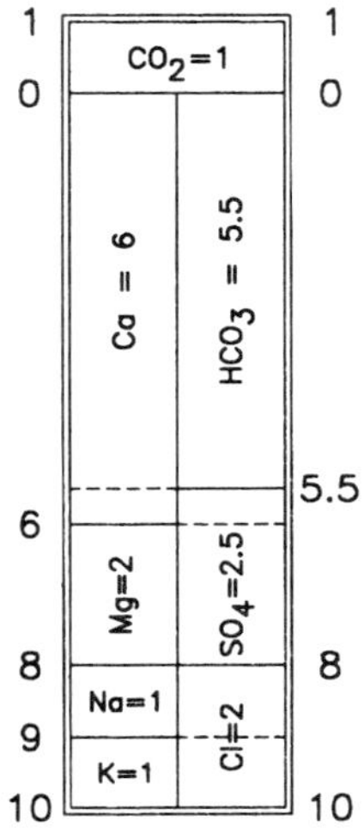


| Hypothetical Compound | Lime (me/L) | Soda Ash (me/L) |
|---|---|---|
| $CO_2$ (1) | 1 | 0 |
| $Ca(HCO_3)_2$ (2) | 5.5 | 0 |
| $CaSO_4$ (3) | 0 | 0.5 |
| $MgSO_4$ (4) | 2 | 2 |
| NaCl (5) | 0 | 0 |
| KCl (6) | 0 | 0 |
| | 8.5 | 2.5 |

FIGURE 10-1. MILLIEQUIVALENT GRAPH OF RAW WATER

(1) Although we do not wish to remove $CO_2$, the lime added for softening will react with the dissolved $CO_2$ according to the following chemical reaction:

$$CO_2 + Ca(OH)_2 \rightarrow CaCO_3 + H_2O$$

Therefore 1.0 me/L of lime must be added to react with the 1.0 me/L of $CO_2$ in the water, producing 1.0 me/L of $CaCO_3$.

(2) 5.5 me/L of lime reacts with 5.5 me/L of $Ca(HCO_3)_2$, reaction 1, producing 5.5 me/L of $CaCO_3$.

(3) 0.5 me/L of soda ash react with 0.5 me/L of $CaSO_4$, reaction 7, producing 0.5 me/L of $CaCO_3$.

(4) 2.0 me/L of lime react with 2.0 me/L of $MgSO_4$, reaction 4, producing 2.0 me/L of $Mg(OH)_2$ and 2.0 me/L of $CaSO_4$. Therefore, 2.0 me/L of soda ash must be added to react with the 2.0 me/L of $CaSO_4$, reaction 7, producing 2.0 me/L of $CaCO_3$.

(5 & 6) Lime and soda ash are not required to react with NaCl and KCl, since these monovalent compounds are not associated with the hardness of water.

The approximate quantities of pure lime and soda ash that must be mixed with the well water to accomplish complete softening are:

**Lime**

8.5 me/L (28 mgCaO/me CaO) = 238 mg CaO/L
or, 8.5 me/L (37 mg $Ca(OH)_2$/me $Ca(OH)_2$) = 315 mg $Ca(OH)_2$/L

**Soda Ash**

2.5 me/L (53 mg $Na_2CO_3$/me $Na_2\ CO_3$) = 133 mg $Na_2CO_3$/L

The **Second Step** is to create the me/L graph that describes the water leaving the settling tanks, and to approximate the quantity of carbon dioxide required to stabilize this water before it passes through the filters. Enough lime and soda ash have been added to fully soften the well water, and therefore, all of the soluble calcium and magnesium compounds have been converted to $CaCO_3$ or $Mg(OH)_2$ precipitates. Since the softening plant is a single-stage facility, we expect 30 mg/L (as $CaCO_3$) of both $CaCO_3$ and $Mg(OH)_2$ to escape the settling process.

30 mg $CaCO_3$/L (me $CaCO_3$/50 mg $CaCO_3$) = 0.6 me/L

Therefore, the me/L graph that describes the water leaving settling is:

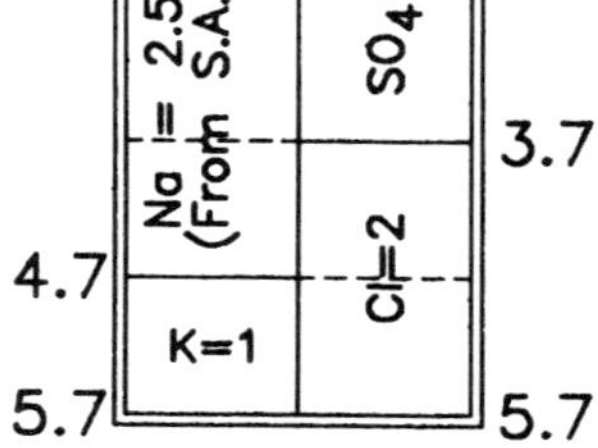


| Hypothetical Compound | | $CO_2$ (me/L) |
|---|---|---|
| $CaCO_3$ | (1) | 0.6 |
| $Mg(OH)_2$ | (2) | 1.2 |
| $Na_2SO_4$ | (3) | 0 |
| $Na_2SO_4$ | (4) | 0 |
| NaCl | (5) | 0 |
| KCl | (6) | 0 |
| | | 1.8 |

FIGURE 10-2. MILLIEQUIVALENT GRAPH OF SETTLED WATER

Note that in the softening reactions, the K, $SO_4^-$ and $Cl^-$ concentrations do not change because these ions are not involved in the precipitates formed in the right hand side of the reactions. However, the $Na^+$ concentration increases because soda ash ($Na_2CO_3$) is added to help form precipitates with the calcium and magnesium hardness compounds.

(1) 0.6 me/L of $CO_2$ react, according to the following chemical reaction, with 0.6 me/L of $CaCO_3$, producing 0.6 me/L of soluble $Ca(HCO_3)_2$ (see Chapter XIII):

$$CaCO_3 + CO_2 \rightarrow\!\!> H_2O + Ca(HCO_3)_2$$

(2) 0.6 me/L of $CO_2$ react, according to the following chemical reaction, with 0.6 me/L of $Mg(OH)_2$, producing 0.6 me/L of slightly soluble $MgCO_3$:

$$Mg(OH)_2 + CO_2 \rightarrow MgCO_3 + H_2O$$

An additional 0.6 me/L of $CO_2$ react, according to the following chemical reaction, with this 0.6 me/L of slightly soluble $MgCO_3$, producing 0.6 me/L of soluble $Mg(HCO_3)_2$:

$$MgCO_3 + CO_2 \rightarrow H_2O + Mg(HCO_3)_2$$

(3,4,5&6) $CO_2$ is not required to react with $Na_2SO_4$, NaCl and KCl, since these compounds are soluble and will not coat the media in the filters.

The approximate quantity of carbon dioxide that must be used to recarbonate the water leaving settling is:

$$1.8 \text{ me/L } (22 \text{ mg } CO_2/\text{me } CO_2) = 39.6 \text{ mg } CO_2/L$$

The **Third Step** is to create the me/L graph that describes the water applied to the filters (i.e., following recarbonation by adding and reacting $CO_2$ with the insoluble precipitates leaving the settling tank). The me/L graph for the recarbonated water is nearly the same as the previous one for the settled water. The differences are the $CaCO_3$ has been converted to soluble $Ca(HCO_3)_2$, and the $Mg(OH)_2$ has been converted to soluble $Mg(HCO_3)_2$.

In this Single-Stage Softening Example, sufficient quantities of lime and soda ash have been added to soften the water completely. The final hardness is:

0.6 me/L Ca + 0.6 me/L Mg = 1.2 me/L hardness

1.2 me hardness/L (50 mg $CaCO_3$/me $CaCO_3$) = 60 mg hardness (as $CaCO_3$)/L

This final hardness is lower than that required in most municipal water systems. Therefore, in many cases less, or no soda ash, and possible less lime can be added so the hardness of the finished water is higher (e.g., 80 to 150 mg/L as $CaCO_3$).

## 2. *Two-Stage, Excess Lime Softening Example*

The flow schematic of a two-stage, excess lime softening facility is:

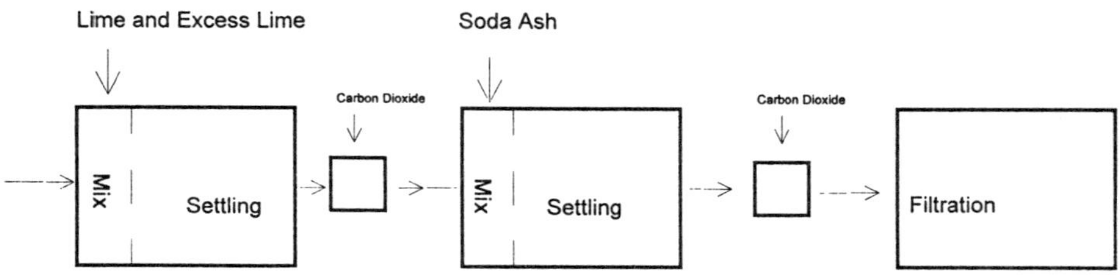


FIGURE 10-3. TWO-STAGE EXCESS LIME FLOW SCHEMATIC

Since the softening reactions can take place in two separate reactors, the first stage can be operated at the optimum condition for magnesium hydroxide removal (pH value of 11.2), and the second stage can be operated at the optimum condition for calcium carbonate removal (pH value of 10.4). Therefore, the level to which the water can be expected to be softened is 10 mg/L (as $CaCO_3$) of magnesium, and 30 mg/L (as $CaCO_3$) for calcium.

The **First Step** is similar to that for the previous example except only the stoichiometric amount of lime required to soften the water, and the excess lime required to raise the pH of the first stage to approximately 11.2, are added in the first stage.

(1) Although we do not wish to remove $CO_2$, the lime added for softening will react with the dissolved $CO_2$. Therefore, 1.0 me/L of lime must be added to react with the 1.0 me/L of $CO_2$, producing 1.0 me/L of $CaCO_3$.

(2) 5.5 me/L of lime react with 5.5 me/L of $Ca(HCO_3)_2$, reaction 1, producing 5.5 me/L of $CaCO_3$.

(3) Soda ash will be added in the second stage to react with the $CaSO_4$ in the raw water.

(4) 2.0 me/L of lime react with 2.0 me/L of $MgSO_4$, reaction 4, producing 2.0 me/L of $Mg(OH)_2$ and 2.0 me/L of $CaSO_4$. Soda ash will be

added in the second stage to react with the $CaSO_4$ formed from this reaction in the first stage.

(5&6) Lime is not required to react with NaCl and KCl, since these compounds are not associated with the hardness of water.

(7) Excess lime is not added for the purpose of reacting with hardness compounds in the water. It is added to raise the pH in the first stage reactor to approximately 11.2. Generally 35 mg/L of CaO are required for this purpose.

35 mg CaO/L (me CaO/28 mg CaO) = 1.25 me/L

The approximate quantity of lime that must be mixed with the raw water to accomplish complete softening, plus raise the pH of the first stage to 11.2, is:

9.75 me/L (28 mg CaO/me CaO) = 273 mg CaO/L
or, 9.75 me/L (37 mg $Ca(OH)_2$/me $Ca(OH)_2$) = 361 mg $Ca(OH)_2$/L

The **Second Step** is to create the me/L graph that describes the water leaving first stage settling, and to estimate the quantity of carbon dioxide required to lower the pH to approximately 10.4 prior to adding

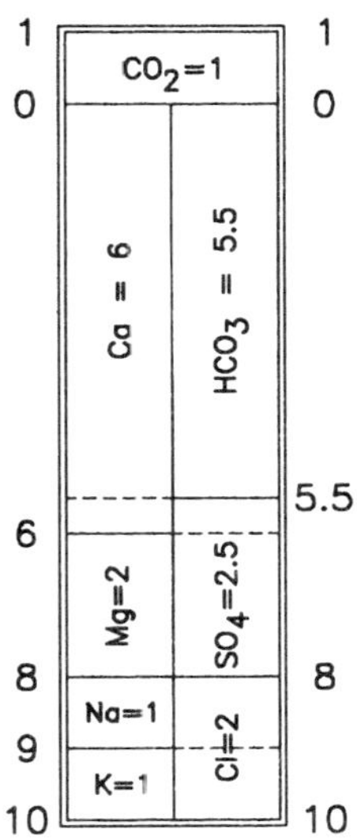


| Hypothetical Compound | Lime (me/L) | |
|---|---|---|
| $CO_2$ (1) | 1 | |
| $Ca(HCO_3)_2$ (2) | 5.5 | |
| $CaSO_4$ (3) | 0 | |
| $MgSO_4$ (4) | 2 | |
| NaCl (5) | 0 | |
| KCl (6) | 0 | |
| | 1.25 | Excess Lime (7) |
| | 9.75 | |

FIGURE 10-4. MILLIEQUIVALENT GRAPH OF RAW WATER

soda ash in the second stage to react with the non-carbonate hardness compounds. Since excess lime has been added to the first stage, we expect 30 mg/L (as $CaCO_3$) of $CaCO_3$ and only 10 mg/L (as $CaCO_3$) of Mg $(OH)_2$ to escape the first stage settling process. Therefore, the me/L graph that describes the water leaving first stage settling is:

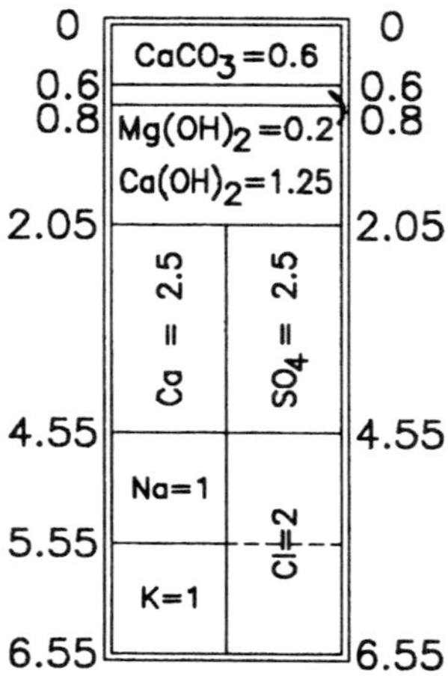


| Hypothetical Compound | $CO_2$ (me/L) |
|---|---|
| $CaCO_3$ (1) | 0 |
| $Mg(OH)_2$ (2) | 0.2 |
| $Ca(OH)_2$ (3) | 1.25 |
| $CaSO_4$ (4) | 0 |
| NaCl (5) | 0 |
| KCl (6) | 0 |
| | 1.45 |

FIGURE 10-5. MILLIEQUIVALENT GRAPH OF FIRST STAGE SETTLED WATER

(1) The $CaCO_3$ leaving the first stage does not have to be carbonated because there is a settling tank in the second stage.

(2) 0.2 me/L of $CO_2$ are added in the first stage to convert the 0.2 me/L of $Mg(OH)_2$ to slightly soluble $MgCO_3$, and to help lower the pH from 11.2 to approximately 10.4.

(3) The 1.25 me/L of excess lime is present since it was not added to react with hardness compounds, but was added to raise the pH to approximately 11.2. 1.25 me/L of $CO_2$ helps lower the pH to 10.4 by reacting with the 1.25 me/L of excess lime, producing 1.25 me/L of $CaCO_3$ that can settle in the second stage.

(4,5&6) $CO_2$ is not required to react with $CaSO_4$, NaCl and KCl.

The approximate quantity of carbon dioxide that must be used to recarbonate the water leaving first stage settling to lower the pH to approximately 10.4 is:

$$1.45 \text{ me/L } (22 \text{ mg } CO_2/\text{me } CO_2) = 32 \text{ mg } CO_2/\text{L}$$

The **Third Step** is to create the me/L graph that describes the water entering the second stage, (i.e., after recarbonation of the first stage effluent), and to approximate the quantity of soda ash required to react with the non-carbonate hardness from the raw water and the non-carbonate hardness compounds formed in the first stage from the lime reactions. The me/L graph that describes the water entering the second stage:

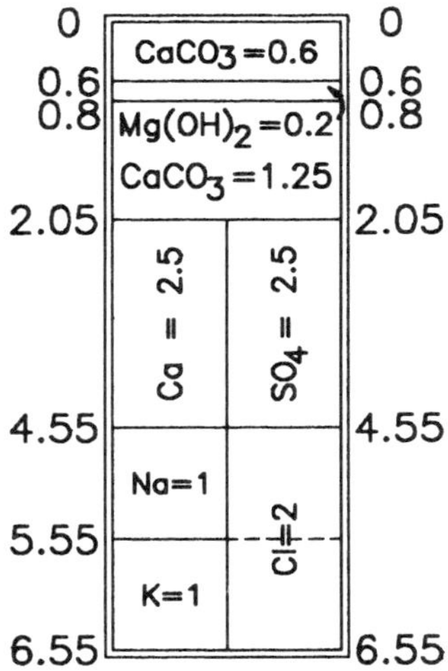


| Hypothetical Compound | | Soda Ash (me/L) |
|---|---|---|
| $CaCO_3$ | (1) | 0 |
| $MgCO_3$ | (2) | 0 |
| $CaCO_3$ | (3) | 0 |
| $CaSO_4$ | (4) | 2.5 |
| NaCl | (5) | 0 |
| KCl | (6) | 0 |
| | | 2.5 |

FIGURE 10-6. MILLIEQUIVALENT GRAPH OF FIRST STAGE RECARBONATED WATER

(1,2&3) Soda ash is not required to react with these precipitates that can settle in the second stage.

(4) 2.5 me/L of soda ash react with 2.5 me/L of $CaSO_4$, reaction 7, producing 2.5 me/L of $CaCO_3$.

(5&6) Soda ash is not required to react with these monovalent compounds, since they are not associated with the hardness of water.

The approximate quantity of soda ash that must be mixed with the water in the second stage to accomplish complete softening is:

$$2.5 \text{ me/L } (53 \text{ mg } Na_2\,CO_3/\text{me } Na_2\,CO_3) = 133 \text{ mg } Na_2\,CO_3/\text{L}$$

The **Fourth Step** is to create the me/L graph that describes the water leaving the second stage settling tanks, and to approximate the quantity of carbon dioxide required to stabilize this water before it passes through the filters. Since the pH of the second stage reactor is roughly 10.4, we expect 30 mg/L (as $CaCO_3$) of $CaCO_3$, and the 10 mg/L (as $CaCO_3$) of $MgCO_3$ from the first stage, to escape the second stage settling process. Therefore, the me/L graph that describes the water leaving the second stage settling is:

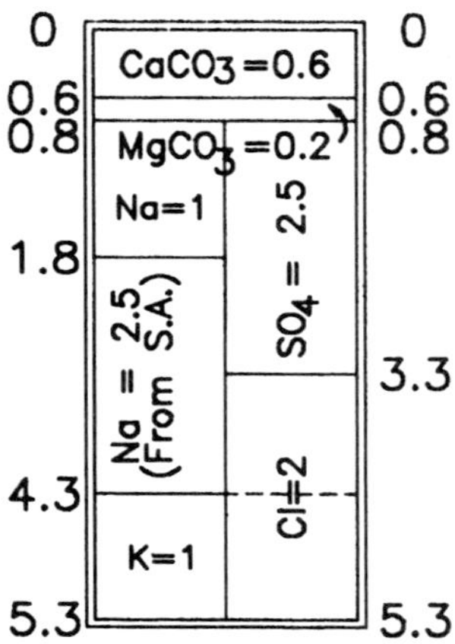


| Hypothetical Compound | $CO_2$ (me/L) |
|---|---|
| $CaCO_3$ (1) | 0.6 |
| $MgCO_3$ (2) | 0.2 |
| $Na_2SO_4$ (3) | 0 |
| NaCl (5) | 0 |
| KCl (6) | 0 |
| | 0.8 |

FIGURE 10-7. MILLIEQUIVALENT GRAPH OF SECOND STAGE SETTLED WATER

(1) 0.6 me/L of $CO_2$ react with 0.6 me/L of $CaCO_3$, producing 0.6 me/L of soluble $Ca(HCO_3)_2$.

(2) 0.2 me/L of $CO_2$ react with 0.2 me/L of the slightly soluble $MgCO_3$, producing 0.2 me/L of soluble $Mg(HCO_3)_2$.

(3,4&5) $CO_2$ is not required to react with these already soluble compounds.

The approximate quantity of carbon dioxide that must be used to recarbonate the water leaving second stage settling is:

$$0.8 \text{ me/L } (22 \text{ mg } CO_2/\text{me } CO_2) = 17.6 \text{ mg } CO_2/\text{L}$$

The **Fifth Step** is to create the me/L graph that describes the water applied to the filters following second stage recarbonation. The me/L graph for the second stage recarbonated water is nearly the same as the previous one for settled water. The differences are the $CaCO_3$ has been converted to soluble $Ca(HCO_3)_2$, and the slightly soluble $MgCO_3$ has been converted to soluble $Mg(HCO_3)_2$.

In this Two-Stage, Excess Lime Softening Example, sufficient quantities of lime, excess lime and soda ash have been added to soften the water completely. Therefore, the final hardness of the finished water is:

0.6 me/L Ca + 0.2 me/L Mg = 0.8 me/L hardness
0.8 me hardness/L (50 mg $CaCO_3$/me $CaCO_3$) = 40 mg hardness (as $CaCO_3$)/L

This final hardness is considerably lower than that required in most municipal water systems. However, this two-stage, excess lime softening example serves a couple of important purposes:

- Often times a two-stage process is necessary to soften surface waters that are hard. One stage is used for softening, and the other stage is used to remove contaminants associated with surface waters (e.g., natural turbidity, taste and odor causing compounds, color producing substances, etc.).
- It also provides an understanding of the calculation procedures for a two-stage process, making it easier to comprehend the more complex split treatment softening process often used in place of two-stage, excess lime softening when the water supply is a ground water source.

### *3. Split Treatment Softening Example*

Split treatment softening is similar to the two-stage, excess lime softening process. However, "excess" lime is not added to raise the pH of the first stage to 11.2. Instead, the total amount of lime required to soften all of the water is added to the first stage, and an excess lime (i.e., high pH) condition is created by passing only a portion of the raw water through the first stage. The hardness of the finished water is higher than that for two-stage, excess lime softening; but split treatment softening of ground water supplies offers several advantages:

1. Excess lime does not have to be added.
2. Carbon dioxide does not have to be added between the first and second stages to recarbonate the excess lime (i.e., to lower the pH from 11.2 to 10.4 for the second stage). The carbon dioxide in the raw water that bypasses the first stage is generally adequate.
3. Less sludge is produced since the excess lime that would be recarbonated to calcium carbonate for settling in the second stage is not needed.

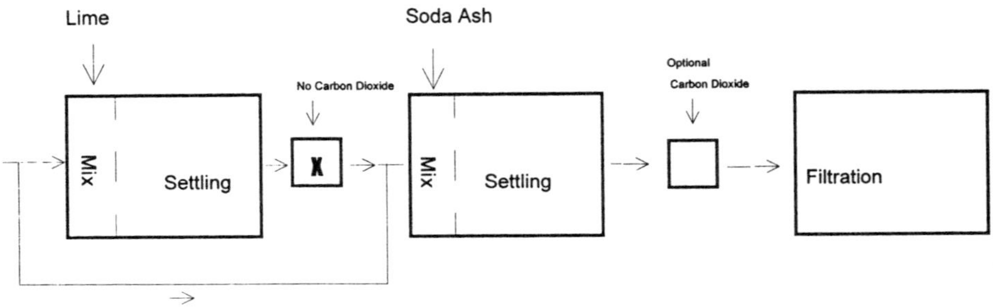


FIGURE 10-8. SPLIT TREATMENT SOFTENING SCHMATIC

The **First Step** is similar to that for the previous two-stage, excess lime softening example, except the pH of 11.2 will be achieved by passing only part of the raw water through the first stage.

(1) 1.0 me/L of lime reacts with the 1.0 me/L of $CO_2$, producing 1.0 me/L of $CaCO_3$.

(2) 5.5 me/L of lime react with 5.5 me/L of $Ca(HCO_3)_2$, reaction 1, producing 5.5 me/L of $CaCO_3$.

(3) Soda ash will be added in the second stage to react with $CaSO_4$.

(4&7) We do not want to produce a finished water with the low hardness that resulted from the previous two-stage, excess lime softening example. A reasonable goal is to leave 1.0 me/L of $MgSO_4$ in the finished water, and to add enough lime to remove only 1.0 me/L of $MgSO_4$. Therefore, 1.0 me/L of lime reacts with 1.0 me/L of $MgSO_4$, reaction 4, producing 1.0 me/L of $Mg(OH)_2$ and 1.0 me/L of $CaSO_4$. Likewise, the second stage soda ash addition will be 1.0 me/L. Therefore, the second stage soda ash required is 1.5 me/L (i.e., 0.5 me/L of $CaSO_4$, formed in the first stage from reaction 4, plus the 1.0 me/L of $MgSO_4$ in the raw water which we have decided to remove).

(5&6) Lime is not required to react with these compounds that are not associated with the hardness of water.

(8) Only enough lime was added to react with 1.0 me/L of $MgSO_4$ in the raw water. Because only a portion of the raw water is passed through the first stage, sufficient lime is available to react with 2.0 me/L of $MgSO_4$. However, even at a reactor pH of 11.2, 0.2 me/L of $Mg(OH)_2$ will escape first stage settling. Therefore, 1.8 me/L (i.e., 2.0 me/L – 0.2 me/L) of lime, or $OH^-$ alkalinity, is actually used up in the first stage softening reactions.

The quantity of lime that must be added to the first stage, based on passing 100 percent of the raw water through the first stage, is:

7.5 me/L (28 mg CaO/me CaO) = 210 mg CaO/L or, 7.5 me/L (37 mg $Ca(OH)_2$/me $Ca(OH)_2$) = 278 mg $Ca(OH)_2$/L

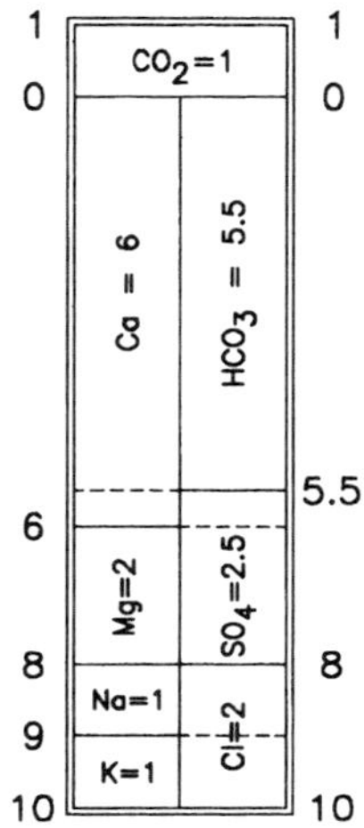


| Hypothetical Compound | Full Flow Lime (me/L) | Lime(OH) Actually Used (me/L) |
|---|---|---|
| $CO_2$ (1) | 1 | 1 |
| $Ca(HCO_3)_2$ (2) | 5.5 | 5.5 |
| $CaSO_4$ (3) | 0 | 0 |
| $MgSO_4$ (4) | 1 (7) | 1.8 (8) |
| NaCl (5) | 0 | 0 |
| KCl (6) | 0 | 0 |
| | 7.5 | 8.3 |

FIGURE 10-9. MILLIEQUIVALENT GRAPH OF RAW WATER.

The **Second Step** is to determine the portion of raw water that should bypass the first stage, and to check whether the excess lime (OH alkalinity) condition created by treating only a portion of the raw water in the first stage is sufficient to raise the pH to 11.2 (i.e., an excess lime concentration of at least 1.25 me/L, or 35 mg/L as CaO). To do this, a mass-balance of the split treatment softening process must be created for magnesium:

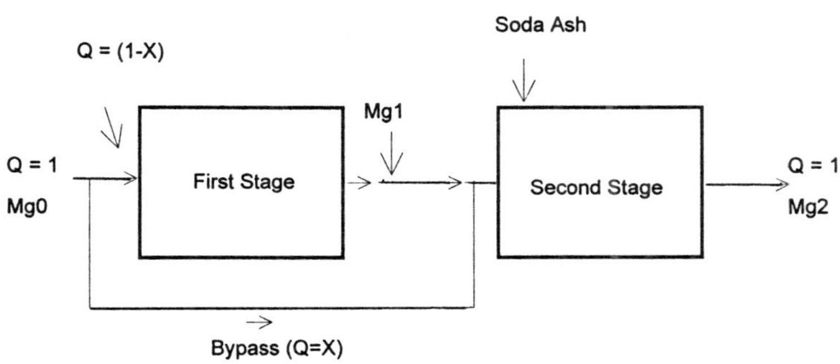


FIGURE 10-10. SPLIT TREATMENT MAGNESIUM BALANCE

Q = 1, where 1 represents 100 percent of the flow
$Mg_0$ represents the magnesium concentration of the raw water
Q = X, where X represents the fraction of raw water bypassing the first stage
Q = (1-X), where (1-X) represents the fraction of raw water to be treated in the first stage
$Mg_1$ represents the magnesium concentration of the flow leaving the first stage
$Mg_2$ represents the goal for the magnesium concentration of the finished water

The Magnesium mass-balance is:

$$Mg_1 \ (1\text{-}X) + Mg_0 \ (X) = Mg_2 \ (1)$$

Rearranging:

$$Mg_1 - (Mg_1 \ (X)) + Mg_0 \ (X) = Mg_2$$
$$(Mg_0 - Mg_1)X = Mg_2 - Mg_1$$

Therefore;

$$X = (Mg_2 - Mg_1) \ / \ (Mg_0 - Mg_1)$$

The goal for the magnesium concentration of the finished water, $Mg_2$, is to leave 1.0 me/L of the $MgSO_4$ from the raw water in the finished water.

$$1.0 \text{ me Mg/L } (50 \text{ mg } CaCO_3/\text{me } CaCO_3) = 50 \text{ mg Mg (as } CaCO_3)/L$$

If sufficient excess lime is available in the first stage, (that is, the pH established is actually 11.2), the magnesium concentration leaving the first stage, Mg1, is expected to be 0.2 me/L.

0.2 me Mg/L (50 mg $CaCO_3$/me $CaCO_3$) = 10 mg Mg (as $CaCO_3$)/L The magnesium concentration of the raw water, $Mg_0$, is 2.0 me/L. 2.0 me Mg/L (50 mg $CaCO_3$/me $CaCO_3$) = 100 mg Mg (as $CaCO_3$). Therefore, the fraction of raw water that should bypass the first stage, X, is:

$$X = (Mg_2 - Mg_1)/\ (Mg_0 - Mg_1) = (50 - 10)/(100 - 10) = 40/90 = 0.44 \text{ (i.e., 44\%)}$$

and, the percent of raw water to be treated in the first stage, (1–X), is:

$$1 - 0.44 = 0.56 \text{ or } (56\%).$$

Now that the percentages of raw water that are to be bypassed and treated in the first stage are known, a check must be made to ensure a sufficient excess lime condition (1.25 me/L) can be created to provide a pH of 11.2:

7.5 me lime/L added, based on treating 100% of the raw water

However, only 56% of the raw water is to be treated in the first stage.

Therefore, the actual concentration of lime in the first stage is: 7.5 me/L (1/0.56) = 13.4 me lime/L

8.3 me/L of lime ($OH^-$ alkalinity) are actually used in the first stage. Therefore, the concentration of excess lime created in the first stage by only treating 56% of the raw water is:

13.4 me/L – 8.3 me/L = 5.1 me/L (OK, greater than 1.25 me/L)

Actually, 0.2 me/L of this 5.1 me/L of excess lime is associated with the $Mg(OH)_2$ precipitate that does not get removed in the first stage, and the other 4.9 me/L (i.e., 5.1 – 0.2 = 4.9) is in the form of excess lime $Ca(OH)_2$.

The **Third Step** is to create the me/L graph that describes the 56% of the water leaving the first stage, and to check whether adequate lime ($OH^-$ alkalinity) will be available in the second stage to react with the raw water that bypasses the first stage. The me/L graph that describes the portion of treated water leaving the first stage is:

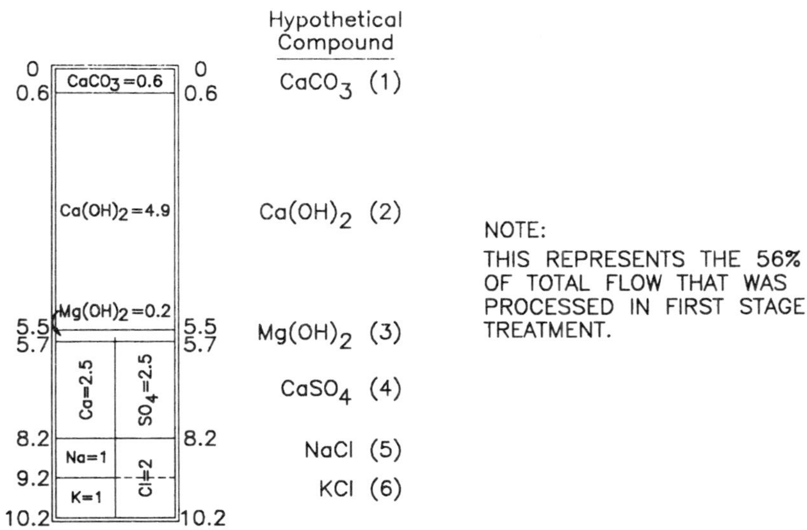


FIGURE 10-11. MILLIEQUIVALENT GRAPH OF 56% FIRST STAGE SETTLED WATER

(1) The 30 mg/L of $CaCO_3$ (0.6 me/L) expected to escape first stage settling.

(2) The excess lime available because only 56% of the raw water is treated in the first stage.

(3) The 10 mg/L, (as $CaCO_3$), of $Mg(OH)_2$ (i.e., 0.2 me/L) expected to escape first stage settling, based on at least 1.25 me/L of excess lime being available in the first stage to reduce the solubility of the $Mg(OH)_2$ to this level.

(4) The 0.5 me/L of $CaSO_4$ associated with the raw water, plus the 2.0 me/L of $CaSO_4$ created from the reaction in the first stage of 2.0 me/L of lime with 2.0 me/L of $MgSO_4$ in the raw water.

The lime, $OH^-$ alkalinity, available in the 56% of water leaving the first stage is:

$0.56\ (4.9\ Ca(OH)_2 + 0.2\ Mg(OH)_2) = 2.86$ me/L

This lime, $OH^-$ alkalinity, is required in the second stage to soften the 44% of raw water bypassing the first stage, by reacting with the $CO_2$ and the $HCO_3$ hardness compounds in this portion of the raw water:

$0.44\ (1.0\ CO_2 + 5.5\ HCO_3) = 2.86$ me/L (this is equal to the equivalent Lime)

The **Fourth Step** is to create the me/L graph that describes the water leaving second stage settling, and to approximate the quantity of carbon dioxide required to stabilize this water before it passes through the filters. The me/L graph of the water leaving second stage settling is:

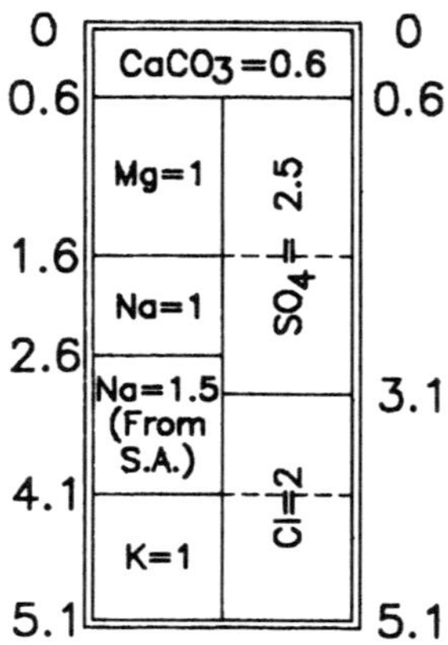


| Hypothetical Compound | $CO_2$ (me/L) |
|---|---|
| $CaCO_3$ (1) | 0.6 |
| $MgSO_4$ (2) | 0 |
| $Na_2SO_4$ (3) | 0 |
| NaCl (5) | 0 |
| KCl (6) | 0 |
| | 0.6 |

FIGURE 10-12. MILLIEQUIVALENT GRAPH OF SECOND STAGE SETTLED WATER

(1) The 30 mg/L, (0.6 me/L), of $CaCO_3$ expected to escape second stage settling. The 0.6 me/L of $CO_2$ reacts with this 0.6 me/L of $CaCO_3$, producing 0.6 me/L of soluble $Ca(HCO_3)_2$.

(2) This is the 1.0 me/L goal established for Mg in the finished water, based on bypassing 44% of the raw water around the first stage and treating 56% of the raw water in the first stage.

$Mg_0$ = 2.0 me/L
$Mg_1$ = 0.2 me/L
$Mg_2$ = 1.0 me/L: i.e., 0.44 (2.0 me/L) + 0.56 (0.2 me/L) = 1.0 me/L

The final total hardness of the finished water from the split treatment example is:

0.6 me/L Ca + 1.0 me/L Mg = 1.6 me/L hardness

1.6 me hardness/L ( 50 mg $CaCO_3$/me $CaCO_3$) = 80 mg hardness (as $CaCO_3$)/L

This is sufficiently soft for most municipal water systems. The four advantages noted at the beginning of this example for split treatment softening should be revisited to better understand the use and benefits of split treatment.

Chapter XI

# SOFTENING WATER FOR MUNICIPALITIES

## Lime and Soda Ash Process

In the most elementary type plant the procedure of treatment is:

1. Addition of softening and coagulating chemicals.
2. Mixing.
3. Settling.
4. Recarbonation.
5. Filtration.

Plants may be continuous or intermittent in operation. The term "continuous type plant" means the continuous application of chemicals during the hours the plant is in operation, and not necessarily that the plant is operated 24 hours a day. Adequate finished water storage, of course, must be provided if the plant is not operated the entire 24-hour period. The cost of a continuous type plant usually is less than the cost of the intermittent or batch type plant (i.e., a plant must be larger if it is to supply the daily finished water demand without being operated 24 hours per day). Almost all municipal plants are the continuous type.

When chemically treated water is mixed and settled in separate tanks the plant usually is referred to as a conventional type plant. However, when the softening chemicals are added so that the softening reactions occur in the presence of a large quantity of previously formed sludge, and clarification is promoted because of upward filtration of the coagulated or softened water through a sludge blanket, the plant is referred to as a sludge blanket type. Many municipal plants of each type have been built.

Formerly, it was not always possible to reduce the hardness of municipal supplies to much under 100 mg/L as $CaCO_3$. However, with the improvements made in recarbonation and the better understanding of the chemical reactions that take place, softer waters can be obtained. It is customary to reduce the carbonate hardness to around 35 to 40 mg/L and the total hardness to approximately 75 to 125 mg/L. To effect economical softening, municipalities usually remove only a part of the non-carbonate hardness. The softened water should be as nearly in chemical balance to calcium carbonate as possible. (See Chapter VIII.)

## Application of Softening Chemicals

Either hydrated lime [$Ca(OH)_2$] or quicklime ($CaO$) may be used in the softening process. The choice depends upon economic factors such as the relative cost per ton of the two materials as well as the size and equipment of the softening plant. Hydrated lime is used more generally in the smaller plants because it keeps better in storage and does not require slaking equipment. On the other hand, quicklime costs less per ton of available calcium oxide and is therefore more economical for use in large plants.

In the early days of water softening it was common practice to add a saturated solution of lime to the water to be softened. It was thought that a true solution of lime could be applied more readily in uniform quantities than milk of lime. The feasibility of thoroughly mixing raw water with milk of lime in large quantities also was questioned. It was believed that the raw water and the milk of lime would not be adequately mixed, resulting in a marked loss in softening efficiency owing to the fact that the suspended particles of lime (calcium hydroxide) would become coated with precipitated calcium carbonate and magnesium hydroxide, thereby becoming inactive.

Experience shows that saturated lime water reacts much more quickly than milk of lime, but is much more difficult to prepare and apply to water than milk of lime. However, The Japanese* and German* water industries are effectively producing large quantities of saturated limewater for both softening and corrosion control. If milk of lime is adequately agitated with the water being softened the results are satisfactory.

Experience at the specific plant and for the actual water being treated is the best way to determine the optimum point of application of chemicals. It is very desirable to have the plant so designed as to permit flexibility of operations at least in this regard. However, it is generally best to add the coagulant, particularly alum, some distance ahead of the lime and soda ash. Lime and soda ash may be added at the same point, but in some cases better results are obtained by adding the lime first, then adding soda ash later. Application of a mixture of lime and soda solutions, while in itself producing good results in many cases, may not be practical because these two solutions, in the concentrations often used, tend to react to form a dense scale of calcium carbonate that not only plugs the solution lines and eductors, but is extremely difficult to remove.

---

* Japanese Municipalities Find Another Way to Add Lime. Sadao Kojima. Water/Engineering and Management, Aug, 1989, page 40.

* Development of a Process for Groundwater Decarbonation with Recovery of High Purity Calcium Carbonate. D. Flinspach. Dr. Eng. Stuttgart State Water Supply Assoc. Feb 27, 1986—Lecture in Muenster, Germany.

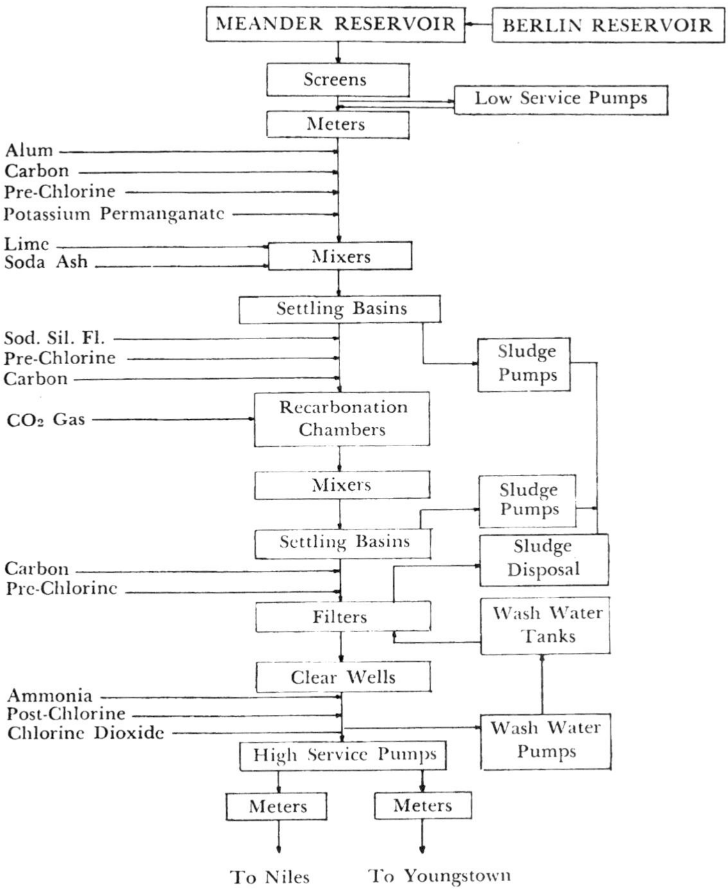


FIGURE 11-1. DIAGRAMMATIC FLOW CHART OF MAHONING VALLEY SANITARY DISTRICT, YOUNGSTOWN, OHIO

The operator of a water softening plant should know exactly how much water is being softened in order to introduce the proper quantities of chemicals. Therefore, the flow of water to the plant should be measured and the chemicals, whether added as a solution or dry, should be fed into the water through a feeding device that increases or decreases its feed in proportion to the flow of the water. If such a device is not provided the water should be delivered to the softening plant through a constant flow regulator. Each time the regulator is changed a proper adjustment should be made on the chemical feeding mechanism.

The continuous feed of standard strength milk of lime is satisfactory. Thorough agitation must be provided to prevent stratification, other-

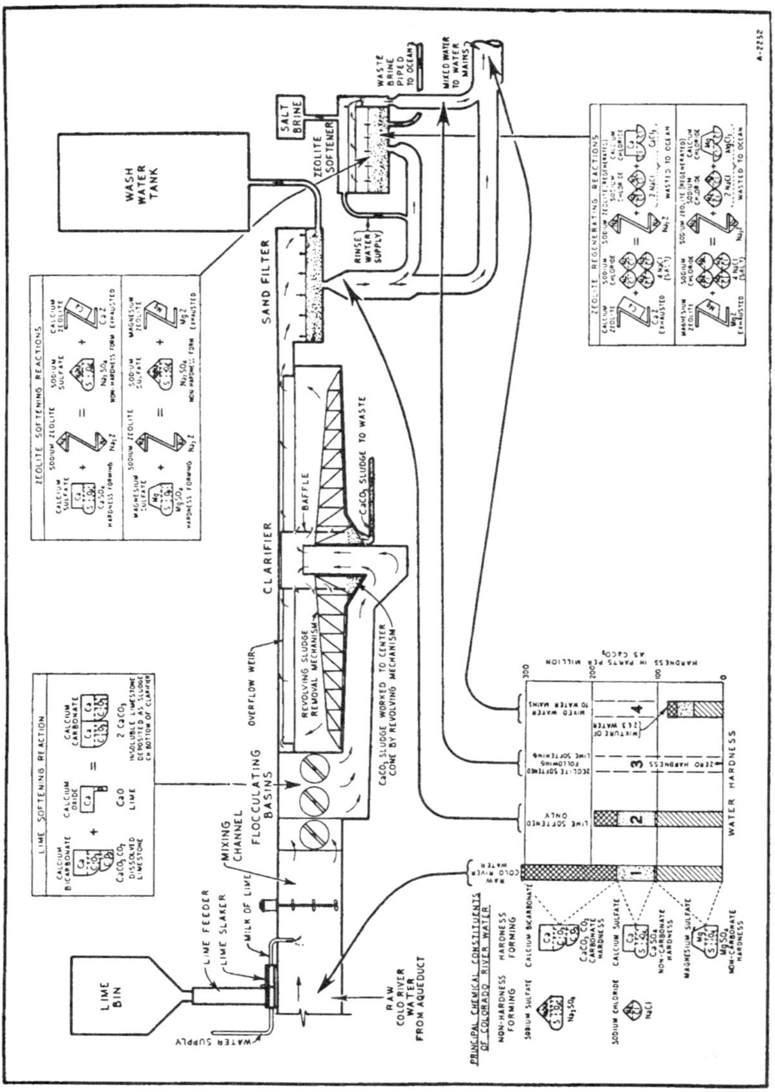


FIGURE 11-2. FLOW DIAGRAM AND SOFTENER REACTIONS, THE METROPOLITAN WATER DISTRICT OF SOUTHERN CALIFORNIA

PICTURE 11-1. MIAMI-DADE

wise the feed of chemicals will not be uniform. The accumulation of fine particles of sand and grit may interfere with the proper operation of the agitator paddles. Due to greater accuracy and simplicity of operation, the use of dry feeding devices for chemicals is becoming more popular in softening plants.

The measurement and feed of chemicals may be controlled either volumetrically or by weight, so that their application to the water will be uniform whether the chemical is coarse, fine, or a mixture of coarse and fine. Machines are available for manual or automatic control. The manual adjustment fixes the quantity of chemical fed according to the rate of pumping. The automatic adjustment, by means of proper connections to a raw water meter, proportions the amount of chemicals fed to the quantity of water flowing to the softening plant.

## Mixing and Settling

Mixing, agitation or flocculation, and settling for the conventional and sludge blanket types already have been described and illustrated. The mixing and settling processes operate in water softening as they do in turbidity removal plants for surface waters, including the concept of the time of mixing and settling in the conventional type plant being

PICTURE 11-2. DAYTON, OH

longer. As explained later, better softening results may be obtained by using a scheme of operation whereby two sets of mixing tanks, settling basins, and recarbonation chambers are used in-series.

### *Overcoming Limitations and Reducing Operating Costs of Process*

Formerly, the reduction of hardness was limited due to the formation of soluble complex basic carbonates of magnesium. These soluble salts caused the alkalinity of the softened water to be high, especially when soda ash was used with the lime to reduce non-carbonate hardness.

The limitations in the reduction of hardness have been largely overcome in the treatment of water for municipal purposes by: excess treatment; excess lime followed by recarbonation; split treatment; excess lime, recarbonation, and secondary stirring, preferably with the addition of water softening sludge or some other finely divided insoluble compound; the use of compounds of aluminum; substitution of the ion exchange process for the addition of soda ash to remove non-carbonate hardness; and the return-sludge process.

*1. Excess Treatment.* This consists in overtreating with lime-usually about 35 mg/L of CaO are sufficient to precipitate the magnesium hardness to approximately 10 mg/L (as $CaCO_3$). The excess lime is then neutralized with soda ash, converting all the causticity to sodium causticity. The soda ash required is that necessary to react with the non-carbonate hardness of the water and the excess lime.

If the water after filtration is recarbonated to pH 8.0, it should be good tasting and have a total hardness of about 40 mg/L (as $CaCO_3$). This is a softer water than is desired in most municipalities. This method of softening water is being used less frequently today because of the advances made with recarbonation systems that provide better protection of the media in the filters receiving the settled water from this single-stage softening process.

*2. Excess Lime Followed by Recarbonation.* This consists in overtreating with lime, as explained above, then neutralizing the excess lime with carbon dioxide ($CO_2$). Carbonate hardness can be reduced by this method to approximately 30 mg/L (as $CaCO_3$).

Excess lime treatment followed by recarbonation can be performed in either a single-stage (i.e., one set of mixing and settling tanks), or in two-stages (i.e., two sets of mixing and settling tanks operated in-series). A two-stage system is often preferred because: the pH can be controlled in the first set of tanks at around 11.2, the optimum value for precipitating magnesium hydroxide; and the pH can be controlled in the second set of tanks at around 10.4, the optimum value for precipitating calcium carbonate; and the excess lime can be recarbonated between the first and second set of settling tanks—providing better protection of the filters that receive water from the second stage.

*3. Split Treatment.* This is a two-stage softening process in which an excess lime condition is established in the first stage by adding the amount of lime required to soften all of the water but only treating a portion of the water to be softened (i.e., some of the raw water is routed directly to the second stage where it is combined with the portion of water treated in the first stage). The $CO_2$ content of the raw water routed around the first stage is used to partially neutralize the excess lime that exits the first stage—reducing the $CO_2$ required for recarbonation. The softened water from a split treatment process typically has a total hardness of approximately 100 mg/L, as $CaCO_3$ (i.e., higher than that associated with excess treatment or excess lime followed by recarbonation, but sufficiently low to be acceptable for most municipalities). Split treatment is particularly useful for softening well water supplies in which the raw water magnesium hardness is less than 40 mg/L (as $CaCO_3$).

**NOTE:** Examples for estimating the amount of lime and soda ash required to soften hard water are provided in Chapter X. In addition to an example for a conventional single-stage softening facility, an example is included for: 1) a two-stage, excess lime facility, and 2) a split treatment facility. The examples in Chapter X should be reviewed to gain a better understanding of the differences between these types of softening plants.

*4. Excess Lime, Recarbonation, and Secondary Stirring.* This procedure is as follows:

(a) Add excess lime and the desired quantity of soda ash.
(b) Mix.
(c) Settle.
(d) Neutralize excess lime with carbon dioxide gas.
(e) Mix, preferably with the addition of some sludge produced from water softening reactions.
(f) Settle.
(g) Recarbonate.
(h) Filter.

For example: A sample of river water having a carbonate hardness of 144 mg/L was mixed with 11 grains of quicklime per gallon and settled. The supernatant water showed the following results of analysis in mg/L:

| | |
|---|---|
| Alkalinity to methyl orange | 78 |
| Alkalinity to phenolphthalein | 61 |
| Excess lime or caustic alkalinity | 44 |

The sample was recarbonated with carbon dioxide gas and then analyzed as follows:

| | |
|---|---|
| Alkalinity to methyl orange | 60 |
| Alkalinity to phenolphthalein | 31 |
| Excess lime or caustic alkalinity | 2 |

Calcium carbonate sludge was added (1500 mg/L), the mixture stirred for less than one-half minute, allowed to settle, filtered through filter paper, and analyzed with the following results:

| | |
|---|---|
| Alkalinity to methyl orange | 15 |
| Alkalinity to phenolphthalein | 4 |
| Excess lime or caustic alkalinity | 0 |

The advantages of the process are:

(a) Less load on the filters.

(b) Less incrustation of sand grains.

(c) A more nearly balanced and therefore a more stable filtered water.

(d) Applicability to the treatment and softening of high turbid water. Mud should be at least partially removed from water before the soften-

PICTURE 11-3. COCOA, FLORIDA SOFTENING PLANT

ing chemicals are added. This can be accomplished in a plant equipped with double mixing and settling units. The procedure for treating flood water is:

(a) Add coagulant and mix.
(b) Settle.
(c) Add softening chemicals and mix.
(d) Settle.
(e) Recarbonate.
(f) Filter.

This process is being used in the many modern softening plants, and is particularly useful when softening hard, surface waters.

*5. Addition of Aluminum Compounds.* Aluminum compounds coagulate colloidal-sized (finely divided) precipitates, thereby facilitating their removal. Without the aluminum treatment, longer reaction periods are required to obtain adequate crystallization of the calcium carbonate and magnesium hydroxide precipitates so they settle properly. The addition of aluminum compounds, according to some authorities, converts soluble magnesium salts to insoluble magnesium aluminates, removable by settling or filtration.

*6. Substitution of Ion Exchange Mineral for Soda Ash to Remove Non-Carbonate Hardness.* Where salt can be obtained at a reasonable cost, it is more economical to remove non-carbonate hardness by means of ion exchange rather than by adding soda ash. The process of operating a lime-ion exchange softening plant is briefly outlined as follows: The water pumped into the plant is treated first with sufficient lime to reduce the carbonate hardness to the lowest possible level. Lime is mixed with the water in the mixing tanks, and the water is then passed through the settling basins or clarifiers to remove the precipitated carbonates. At the end of the settling basins the lime softened water is recarbonated with carbon dioxide gas to convert the normal carbonates to bicarbonates, thus preventing the deposition of normal carbonates on the ion exchange medium and the formation of sodium carbonate in the softened water. From the settling basins the water flows to filters. If it is desirable to reduce it to zero hardness, all the water is passed through ion exchange softeners. The Metropolitan Water District of Southern California is the largest plant in the world using this process (See Fig. 11-2).

*7. Return Sludge in Lime-Soda Ash Water Softening.* It has been found that returning part of the sludge formed in the softening basins to the head of the mixing tank and mixing it again with the water being treat-

PICTURE 11-4. LAGOONS AT SOFTENING FACILITY

ed improves the chemical reactions taking place. The benefits derived are a slight reduction in the carbonate hardness and the more complete stabilization of the water. The benefits are probably due to "mass action" and the presence of unreacted lime in the sludge. Concentrations to be carried may vary with the type of water being treated. Some plants in successful operation use the following test: 100 ml. sample from the mixing tank effluent is measured into a graduate cylinder. After settling 30 minutes, the sludge should have a volume of at least 5 ml.

## Lime Recovery

In most cases it is neither economical nor practicable to reclaim the lime from the calcium sludge by processing it and reburning it. Where it is feasible to do it, a two-stage softening process is generally used to soften the water. Sufficient lime to precipitate only the calcium and the required amount of soda ash are added in the first mixing chamber. Sufficient lime to reduce the magnesium hardness is then added in the second mixing chamber. The precipitated sludge in the first settling basin is largely calcium carbonate, which can be dried, recalcined and used as a source of lime for softening. The precipitated sludge in the second settling basin is largely magnesium hydroxide, which must be separated from the calcium carbonate to prevent progressive increase in the inert material of the lime produced from the sludge. In waters which are extremely low in magnesium content, it may be possible to reclaim the lime from all of the sludge.

In consideration of lime recovery (or recalcination) as compared to various means of sludge disposal, a complete economic study should be made. Recalcination, except in a few situations, may be more costly than disposal if the complete costs are all taken into account.

Softening plants such as the Dayton, Ohio plant and the Miami-Dade (Florida) Hialeah (Pic. 11-1 & 11-2) plant are among the largest recalcining plants in the United States. The Hialeah plant produces thirty percent more quicklime by recalcining than is added in the softening process. It has the capacity to produce 100 tons of quicklime per day.

One patented process separates the magnesium from the calcium by "scrubbing" the stack gases (from the recalcining furnace) through the slurry. The $CO_2$ dissolves almost all the magnesium hydroxide converting it to soluble magnesium carbonate before reacting with the calcium carbonate. This process requires careful control to avoid converting the insoluble calcium carbonate to soluble calcium bicarbonate.

In the case of softening turbid surface waters, the lime cannot be reburned from the sludge unless another step of coagulation and sedimentation is used ahead of the mixing and settling process described above. Considerable research remains to be done to make it feasible to recover a fairly pure lime from the sludge that results from clarifying and softening surface water in the same operation.

The recovery of lime from softening sludges is practiced in the U.S. today in less than a dozen water treatment plants. Generally, for recalcination to be practicable: 1) the magnesium content of the sludge should be low, 2) the sludge should be dewatered to a solids content of 70% or higher before it is burned, and 3) a lime production rate of at least 50 tons/day is desirable.

## Sludge Disposal/Beneficial Reuse

In the past, water softening plants located along larger streams often discharged the sludge into these waterways. Current stream pollution regulations have forced a stop to this practice so that other means of disposal must be used. All water softening plants, large or small, are now being required to refrain from directly discharging any wastes, liquid or solid, into the rivers and streams.

Where abandoned stone quarries, sand and gravel pits, or coal mines are available nearby, it may be possible to dispose of the sludge for many years without any serious difficulty. In areas having abandoned coal-strip mines, pumping the sludge into this area may be the logical solution and most fortuitous. If acid waters are leaching out of the mine into the watershed, as many such mines do, the waste sludge can help in abating serious stream pollution by neutralizing the acid water. Such neutralization offers a great potential savings in material treatment cost to the state.

Lagooning is another practical solution. If the sludge can be dried to about 50% moisture content in the lagoons, the annual requirement will be about 0.5 to 1.0 acre-feet per 100 mg/L hardness removed per MGD.* Also, multiple lagoons used on a fill and let dry sequence, or filling from one end and withdrawing or decanting from the other, often result in better drying of the sludge.

The least costly method of transporting sludge to a lagoon is by pumping through a pipeline. Other plants, to facilitate handling of the sludge for discharge into lagoons or pits, use mechanical equipment (e.g., centrifuges, belt filter presses, plate and frame presses, vacuum-assisted drying beds, vacuum filters, etc.) at the water plant to thicken and/or dewater the sludge into a concentrated solid form. With such equipment it is possible to reduce the moisture content to about 50%, and at this level the sludge is considered acceptable land-fill, offering another solution.

Increasing ingenuity is necessary for sludge disposal since lagooning, one of the most common practices, is becoming more difficult due to the dearth of available land and/or soaring land costs. Therefore, other methods are being increasingly employed.** Discharging the sludge directly into a city's sewage system is being practiced in some cases. Such a discharge must be done with the approval of the municipal sewage department since not all systems can accommodate this alkaline waste, at least in the volume available. In other instances the discharge of such sludge could prove beneficial in neutralizing other acid trade wastes entering the same system. A few small plants are permitted to discharge directly into the city's storm sewer system, although this method has been phased out in most cases.

An attractive method for a water plant located in a farming area, is to use land application of the sludge on farm fields. Lime softening sludge is generally a pure source of calcium carbonate, with varying small amounts of magnesium hydroxide, and provides an effective source of liming material for farmers. It has a neutralizing value for soil acids in excess of most agricultural liming materials. As a result, some utilization is made of the sludge for this purpose. But due to the fact

* Howson, L. R., Journal, A.W.W.A., Vol. 53, p. 1169, Sept. 1961.

** For additional information, see the A.W.W.A. Research Foundation Reports, "Water Treatment Plant Waste Management," June 1987; and "Slib, Schlamm, Sludge," 1990.

PICTURE 11-5. SLUDGE SPREADERS.

that sludge as normally produced does not dry out sufficiently to be handled easily, farmers are reluctant to go to the trouble required to move the material. However, spreaders have been developed for handling wet sludge, which are mounted on a big-wheeled truck; the wide flotation tires allow spreading of the wet material on soft ground without bogging down (see photo). The more advanced spreading systems developed for wet material should aid water plants in disposing of sludge on farm fields at reduced cost or even slight profit.

Some water plants also thicken and/or dewater the sludge with the mechanical equipment noted previously followed by air drying in outdoor stockpiles. Some plants using this practice are reported to have a demand for the dried sludge that is greater than the amount available.

Lastly, the Spiractor equipment for lime-soda water softening, described in Chapter III, offers an intriguing solution to sludge disposal. Instead of wet sludge, it yields a dry, hard, granular calcium carbonate. Although easy to handle and satisfactory to use as landfill, it potentiates value as a salable by-product. Spherical in shape, these light to white-colored granules of 1/4 in. to #8 mesh size are used by home gardeners as surface material for decorative effects in the garden and walks. Since it is derived from micron particles of precipitated calcium carbonate, it would be much more reactive in neutralizing acid wastes than limestone of the same or smaller size and chemically purer. With ingenuity other beneficial uses could be developed.

## Chapter XII

# Removal of Iron and Manganese, Heavy Metals, and Other Undesirable Constituents

Iron is present in many ground water supplies whereas manganese is present only occasionally and then usually along with iron. In many states iron is present in ground waters rather generally. In Ohio, for example, dissolved iron (Fe) in well waters ranges from a few hundredths to 77 mg/L.

The presence in a water supply of over 0.3 mg/L of iron, or iron and manganese, is considered objectionable. The amount of manganese should be low, otherwise, troubles will result even with the combined amount being under the 0.3 mg/L level. Manganese of even 0.01 or 0.02 mg/L can be very objectionable. The current Safe Drinking Water Act (SDWA) Secondary Maximum Contaminant Levels (SMCLs) for iron and manganese are 0.3 and 0.05 mg/L respectively. These compounds are objectionable because they stain everything with which they come in contact. This is especially true of plumbing fixtures and of clothes that are rinsed in the water.

Iron-bearing water sometimes is so badly discolored that it is not pleasant to use. In extreme cases, iron and manganese compounds in water interfere with culinary uses such as causing tea to turn black and darkening vegetables cooked by boiling. Iron-bearing waters do not always behave satisfactorily in the distribution system because they stimulate the growth of Crenothrix and other bacteria which may impart disagreeable tastes and odors to the water. The crenothrix growths are stringy and gelatinous and may become so voluminous as to interfere with the flow of the water through the pipes. Chlorine treatment sometimes is used to prevent their growth.

One of the greatest annoyances of iron in water is that it precipitates out of solution and settles in the mains, especially in the outlying districts where the velocity of flow is low. Sudden demands for extra water, such as the opening of a fire hydrants, stir up this rust deposit to such an extent that the consumer has great difficulty in using the water from the mains. Manganese rust or stain, especially on clothes, is even more objectionable than iron rust or stain because it is darker. Deposits of manganese oxide are almost black.

Iron-bearing water when first pumped from the ground usually is clear because the iron is present as ferrous iron, and in that condition is soluble. Water may be perfectly clear, however, and still have several hundred mg/L of iron dissolved in it. As soon as it absorbs oxygen, the iron is changed from the ferrous to the ferric state and forms solids which precipitate completely or exist as a colloid. Iron in the colloidal condition causes the water to be badly discolored. A sample of water containing iron may be perfectly clear when first pumped from a well, but in a few hours it will change to a brown color, or perhaps contain a brown sediment.

If public water supplies contain more than 0.3 mg/L of iron, it is considered good practice to remove it, or perhaps sequester it with sodium polyphosphate, unless there is also present certain organic humus material that prevents precipitation of the iron.

Iron and manganese, whether existing in the soluble or the insoluble state, exhibit limited presence in filtered water, and therefore their removal is incidental to the normal processes of coagulation and flocculation. However, other pretreatment techniques are usually necessary for the removal of these two metals, and the choice of technique is dependent on the source water and the forms of iron and manganese present.

The lime softening process is excellent for the removal of iron and manganese, whether in the soluble or insoluble state. This process is very effective in removing the difficult to remove manganese, because of the high pH in the reaction process. This is particularly true of the iron and manganese found in surface supplies.

Iron and manganese found in groundwater exists in one of three states: the Ferrous ion, $Fe^{2+}$; the Manganous ion $Mn^{2+}$; and the complexed forms which are associated with color (i.e., iron or manganese complexed with organic material). For the latter, removal may be dependent upon aeration followed by filtration, or oxidation by chlorination or ozonation followed by filtration, or by simple filtration.

The ferrous and manganous ions can be removed with hardness by ion exchange (e.g., zeolite) processes. Care must be taken to avoid air contamination of the feedwater, or resin fouling can result. Specially maintained manganese-zeolite exhangers, in which $KMnO_4$ is used to coat the exchange media, are very effective for the removal of manganous ion.

Common filter media (e.g., anthracite coal and silica sand) are used in many water plants to remove soluble manganese ($Mn^{2+}$) ions from water. In these cases preoxidation of the $Mn^{2+}$ well ahead of the filters is not practiced. As the water is applied to the filter, an oxide coating ($MnO_2$) is formed on the media. The $Mn^{2+}$ ions are adsorbed onto the oxide coated media, usually in the upper 3 in. of the filter bed. Chlorine is often applied just prior to the filter, and serves to oxidize the $Mn^{2+}$ adsorbed on the media. Manganese removed in this way is retained on the media surface and is not significantly removed during backwash of the filter. For this process to work effectively the pH of the water should be basic. Also, oxidants stronger than chlorine are not generally applied just prior to the filter because they tend to oxidize the $Mn^{2+}$ instantaneously, as opposed to oxidizing the $Mn^{2+}$ after it has been adsorbed on the oxide coating of the media.

Where ferrous and manganous ions are found in carbonate-bearing well water, the addition of lime can effectively and completely precipitate these ions as iron and manganese carbonate when the pH is kept at 8 to 8.5. If the pH is allowed to rise to 11 with lime addition, the iron and manganese precipitate in the hydroxide form: $Fe(OH)_2$ and $Mn(OH)_2$.

## Hydrogen Sulfide

Hydrogen sulfide is about twice as soluble as carbon dioxide, and when dissolved in water, produces an offensive odor resembling that of rotten eggs. It is sometimes present in deep well water due to reduction of sulfate. It can be reduced by aeration, and that which is not removed is quickly oxidized to free sulfur by the dissolved oxygen in the aerated water. The precipitated sulfur should be removed by filtration because it may be reverted back to hydrogen sulfide by the action of certain microorganisms. It can be removed also by chlorination. Enough chlorine must be used to maintain a distinct chlorine residual.

The removal of hydrogen sulfide by chlorine, rather than by aeration, may be advantageous in cases where it is not necessary to repump the water because chlorine can be fed directly into the main. The reaction between chlorine and hydrogen sulfide is as follows:

$H_2S + Cl_2 \rightarrow 2HCl + S$. For the removal of each pound of hydrogen sulfide 2.1 pounds of chlorine are required.

## Fluorides

The presence of over 1.2 mg/L of fluoride (or less in waters of high average temperatures) is not desirable; and over 2 mg/L is dangerous, since this amount in water used for cooking and drinking will cause mottling of tooth enamel, especially during the period of tooth development in children. (See Chapter VI).

## Removal of Fluorides.

Lime softening can be used to lower the fluoride content of the water being treated, and the extent of reduction is related to the magnesium removal by the formula:

> Fluoride reduction equals the original fluoride concentration in mg/L × .07 × the square root of the magnesium removed.

Ordinary ion exchange softening does not remove monovalent fluorides, but some types of exchange material have been found to absorb fluoride after the softening cycle is completed. The fluoride is then unloaded after regeneration, during the start of the next softening cycle. This probably accounts for the fluctuations in the fluoride content of some water supplies.

At Bartlett, Texas where the raw water supply contains 8 mg/L fluorides, they are removed by the activated aluminum process. Other materials that have been considered for fluoride removal are: aluminum sulfate, tricalcium phosphate, calcium phosphate, and ground bone.

In addition to the previously mentioned metals and fluorides, inorganic contaminants may be present in the potable supply. The overall concentration of the inorganics may result from naturally occurring minerals in the source water, from man made pollution, or by-products of the corrosivity of the potable supply. Contaminants such as arsenic, lead, copper, barium, cadmium, chromium, mercury, selenium and silver, as well as specific ionic salts of chlorides, nitrates and fluorides may be present, and all are subject to limitation by established maximum contaminant levels as permitted by the Safe Drinking Water Act (SDWA).

No cost effective treatment method exists singly whereby all of these contaminants can be removed. If a source water cannot be found which is free from one or more of these constituents, treatment becomes necessary. Further, a corrosive water may leave the treatment plant free of contamination, but pick up traces of metals in the home plumbing system in sufficient quantity to cause concern. As a result, certain special practices have developed which have enabled water plant operators to cope with the situations.

Distillation of water is the most effective method of removing all inorganic contamination from water. It's high energy requirements, however, restrict the use of this technique to the laboratory. Conventional water treatment schemes are employed to remove many of the contaminants with excellent results. Coagulation with aluminum or iron salts, and softening and chemical precipitation with lime will remove many heavy metals in source water.

Arsenic removal in the range of 90% or more can be achieved with normal coagulation using Alum or Ferric salts. Alternatively, lime softening can achieve equal removal percentages of Arsenic concentrations.

Barium, cadmium, chromium, radium and silver are readily removed by the lime softening process, but only to a lesser extent than is achieved by the conventional coagulation methods with alum or ferric. (Barium is not removed by alum or ferric).

Mercury is not readily removed by lime treatment of water supply, and is only partially removed by conventional alum and ferric coagulation. Lead is easily removed by lime softening.

As mentioned in an earlier chapter, ion-exchange is used in many water plants for the removal of hardness, iron and manganese, and for specific ion removal. Refer to the chapter on ion-exchange for reference.

## Reverse Osmosis (RO)

Salt water intrusion of source supply, pollution of sources, the establishment or organic complexes as priority pollutants, and the need for point of use (POU) home devices has brought the process of reverse osmosis into vogue. Osmosis is the spontaneous passage of a liquid through a membrane from a dilute solution to the more concentrated one. This membrane, through which passage of water occurs, holds back any dissolved solids present, and is therefore called semi-permeable.

Water, as a compound, possesses the natural property of osmotic pressure, which is a function of the concentration of any salts which may be dissolved in it. When a semi- permeable membrane is placed between a salt solution of low concentration and a salt solution of high concentration, water leaves the area of lower concentration to migrate through the membrane to the area of higher salt concentration.

Reverse Osmosis, then, seeks to promote the opposite migration. Given the same two salt solutions as described above (with membrane), a pressure sufficient to overcome the natural osmotic pressure is applied to the highly concentrated salt solution. This brings osmosis to

PICTURE 12.1. DARE COUNTY REVERSE OSMOSIS

a standstill. As more pressure is applied, water begins to migrate from the higher concentrated salt solution, leaving a brine or brackish water, while producing a water of low or no salt concentration on the other side of the membrane. While this idea is not new, recent state-of-the-art advances have allowed the application of this technology to the water supply field.

Perhaps one of the most ambitious applications of this process can be found at the regional water supply system operated by Dare County, North Carolina, which serves residents in and around Nags Head and Kill Devil Hills on the Outer Banks of the Atlantic Ocean. Here, a 3 MGD RO facility was designed and made operational in August of 1989. This $7.5 million facility treats raw well water containing an average of 3,200 mg/L total dissolved solids (TDS), and is able to meet the requirements of the SMCL for TDS of 500 mg/L.

The source of supply for this plant is eight wells which have been drilled to a depth of 325 to 425 feet. Rated at 500 gpm each, the wells provide water which is high in hardness (350mg/l as $CaCO_3$), high in sodium chloride (in excess of 800 mg/L), and has a conductivity that approaches 6,500 micro- mhos per centimeter. As this brackish water is drawn from the aquifer, ocean water intrusion results in an ever increasing salt content in the source. Diligent operation of the processes in this well run facility serves to minimize this problem, as operators achieve a 75% recovery rate in their system.

Well water is treated with sulfuric acid and scale inhibitor to discourage calcium carbonate fouling of the delicate RO membranes. For additional protection, cartridge filters are used to remove sand and other particulate contamination down to 5 microns. The resultant effluent is called feed water, and it is sent to the treatment racks by three 250 horsepower variable speed driven vertical turbine pumps which operate at 275 psi. When operated properly, these two- pass racks are guaranteed to produce excellent quality water for five years.

Waste from these racks is concentrated brine, which is pumped back to the ocean under a NPDES permit. The salt concentration of this waste is only 1/3 that of sea water. The racks produce a product water (permeate) which is characterized by a TDS concentration of only 75mg/l. To prevent the inevitable corrosivity of such a water, it is blended with disinfected well water in such proportion that a plant effluent of 375 to 400 mg/L TDS is achieved. This blending produces a higher plant capacity, and lower cost per thousand gallons treated (approximately $1.16/1000 gals which includes salaries, supplies, etc.).

Before the water is pumped to the distribution system, it is treated with 25% NaOH to bring the pH to above 8, and with Calgon C-10 Zinc Orthophosphate for corrosion control. Chlorination and planned fluoridation round out the treatment scheme.

Although RO is the membrane process that has been used most extensively in water treatment, other membrane processes are also available. RO is used mainly for the treatment of brackish waters, and pressures required range from 250 psi up to 1,000 psi. In addition, other membrane processes are being used to remove undesirable contaminants from water supplies. Nanofiltration membranes, operated at pressures of 75 to 250 psi, are being used to soften waters. Ultrafiltration and microfiltration, which can be operated at even lower pressures, are also being considered for water treatment. Membranes are no longer being considered only to lower the TDS concentration of a water. More stringent drinking water regulations are making membrane processes more attractive for removing a variety of contaminants (e.g., microorganisms, color-producing substances, organics, turbidity, etc.).

Chapter XIII

# Recarbonation of Lime Softened Water

Lime softened water which has not been recarbonated is not stable because it is supersaturated with normal carbonates of calcium and magnesium. Distribution systems carrying the water can become incrusted with deposits of these carbonates. Difficulty with the deposits is encountered also in meters and hot water systems, especially in the hot water pipes of hotels and public buildings because the water in these buildings usually is heated to a temperature higher than the average. See accompanying photographs 13-1 and 13-2.

Unstabilized lime softened water can have an injurious effect on sand filters. The normal carbonates deposit on the surface of the sand grains as a crystalline coating and cause an enlargement of the grains. The effective size and uniformity coefficient of the sand grains change as the crystalline coating increases, affecting the ability of the filter to operate properly. The sand grains can also become cemented together, forming hard lumps in the filters and eventually destroying the effectiveness of the filter medium. (See photograph in filtration chapter). As the sand grains increase in size, the filter bed increases in depth, making it necessary to periodically remove some of the sand in order to maintain the proper freeboard between the top of the sand and the wash water troughs. Also it may be necessary to occasionally spade up the bed and crush the hard lumps. After four to ten years the sand medium may have to be replaced and the filters may require rebuilding.

Lime softened water can be made stable by recarbonating it with carbon dioxide gas. The gas is diffused into the water through small openings in a grid system. The carbon dioxide neutralizes any excess lime or caustic alkalinity in the water, and if added in sufficient quantity converts the normal carbonates to bicarbonates. It is usually applied after the water has passed through the settling basins and before reaching the filters. It is preferable, however, to provide some reaction time for the recarbonated water before it reaches the filters. Typically recarbonation basins are provided with an overall detention time of approximately 20 minutes. The basin is usually divided into two compartments. The first compartment, which is used to disperse carbon dioxide into the water, has a detention time of roughly three minutes. The second compartment provides the remainder of the detention time to allow the carbon

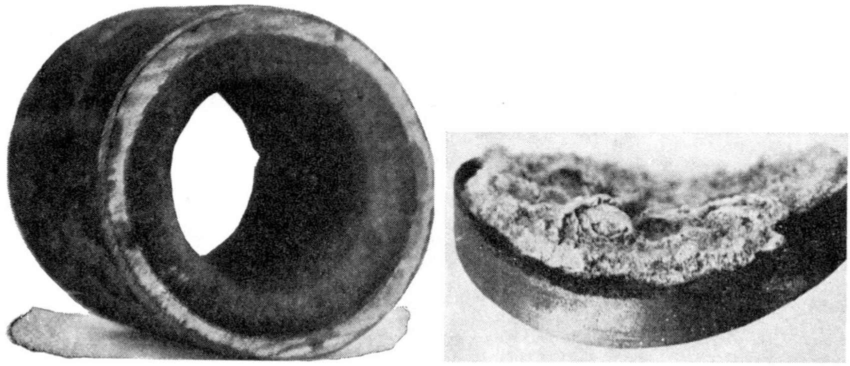

PICTURE 13-1. LIME SCALE

dioxide to completely react chemically with the water. A baffle is normally provided to separate the two compartments. Water from the first compartment, in which the carbon dioxide is introduced through diffusers, is forced to flow under the baffle into the second compartment.

In the time since recarbonation has come into general practical use, the conception of its function has changed almost completely. The idea during the development of the process was to add enough carbon dioxide to the softened, settled water to neutralize any excess lime or caustic alkalinity in the water, and to convert all of the normal carbonates to bicarbonates (i.e., to recarbonate to a pH value of about 8.3). When this was tried on a plant scale, however, it was found that the low alkalinity softened water had an aggressive reaction on deposited scale, on the precipitates already accumulated in the basins, and on the incrusted filter sand. In fact, the water after recarbonation would dissolve scale and precipitates to such an extent that the alkalinity in many cases would be more than doubled. In the light of present knowledge of calcium carbonate saturation, the increase in alkalinity can be explained

PICTURE 13-2. LIME SCALE

readily, and easily computed by means of the nomograph of the Langelier formula. For example, to be in chemical balance a sample of softened recarbonated water having an alkalinity of 35 mg/L, total solids of 300 mg/L, and calcium of 20 mg/L should have a pH of 8.9. To put this water in chemical balance at a pH of 8.3 the alkalinity would have to be about 130 mg/L, or almost four times the alkalinity required for a pH of 8.9.

Another difficulty in recarbonating the alkalinity of lime softened, settled water to bicarbonates is the tendency to return the finely suspended calcium carbonate and magnesium hydroxide into solution. For example, if a sample of softened, settled water is divided into two portions, and the alkalinity of one portion is recarbonated to all bicarbonates, the alkalinity after filtering both portions would be higher in the recarbonated sample. It is entirely probable that this increase in alkalinity is due more to the dissolving of the magnesium hydroxide than the calcium carbonate. This is because magnesium hydroxide can be dissolved very readily by carbon dioxide, whereas calcium carbonate is much more difficult to put into solution.

Lime softened water with an analysis for alkalinity of 40 mg/L or under, a hardness of 85 mg/L or under, and a calcium of 18 mg/L or less, must have a pH value of about 8.9 to be in equilibrium. Keeping in mind that low alkalinity, oxygen-bearing water should have a stability index of +0.4 to +0.8 to prevent red water troubles, it is apparent that recarbonation of low alkalinity water should be performed to lower the pH value only in the range of 9.3 to 9.7. This creates a water that tends to incrust the sand filters. However, it is believed to be a better policy to keep the trouble in the plant, than to distribute a water that could be aggressive to the consumers' plumbing. If the stability index of the water is too positive, and the water tends to lay down excessive scale in the cold water pipes, the scaling can be reduced by adding a trace of sodium poly-phosphate (0.25 to 1.0 mg/L).

## Recarbonation not Always Necessary

Experiences with both well supplies and moderately hard surface supplies have produced some unexpected results. Rather than operate the recarbonation equipment, the sand filters have been permitted in some cases to become coated with calcium carbonate, producing an effluent that is reasonably stable. Even at pH values of 10.5 or so, deposition in these systems has been slight and after several years this practice has not generated any unusual complaints. The additional costs of filter maintenance in these systems have been found to be more justifiable than the maintenance of recarbonation equipment. In none of these cases is there any additional stabilizing treatment with polyphosphate.

## Calculation of Carbon Dioxide Requirements

Carbon dioxide is used mainly in three types of chemical reactions to convert alkalinity from one form to another:

1. To convert caustic (hydroxide) alkalinity to carbonate alkalinity.

   a. $Ca(OH)_2 + CO_2 \rightarrow CaCO_3 + H_2O$ (1a)
   b. $Mg(OH)_2 + CO_2 \rightarrow MgCO_3 + H_2O$ (1b)

2. To convert carbonate alkalinity to bicarbonate alkalinity.

   a. $CaCO_3 + CO_2 + H_2O \rightarrow Ca(HCO_3)_2$ (2a)
   b. $MgCO_3 + CO_2 + H_2O \rightarrow Mg(HCO_3)_2$ (2b)

3. To convert caustic alkalinity to bicarbonate alkalinity (i.e., chemical reactions 1a and 2a, or reactions 1b and 2b).

The carbon dioxide required for recarbonation can be estimated through calculations that take into account the chemical reactions listed above. The calculations can be performed on a molar basis if a balanced chemical reaction is used (i.e., the moles of each reactant are known). An alternative is to perform the calculations on an equivalent basis, realizing that for a single chemical reaction one equivalent of any reactant reacts with one equivalent of another reactant. The following examples illustrate the use of both the molar and equivalent means of estimating carbon dioxide requirements. Since it is standard practice to express all forms of alkalinity as $CaCO_3$, the molecular weight (100 g/mole, or 100 mg/m mole) and equivalent weight (50 g/eq, or 500 mg/meq) of $CaCO_3$ are used in these examples.

**1. Caustic alkalinity to carbonate (using chemical reaction 1a)**

a) molar basis

1 mole of $CO_2$ reacts with 1 mole of $Ca(OH)_2$, as $CaCO_3$; or, 1 m mole of $CO_2$ reacts with 1 m mole of $Ca(OH)_2$, as $CaCO_3$

1 m mole $CO_2$/1 m mole $CaCO_3$ × 44 mg $CO_2$/m mole $CO_2$ × m mole $CaCO_3$/100 mg $CaCO_3$ = 0.44 mg $CO_2$/mg $CaCO_3$

If the concentration of $Ca(OH)_2$ (mg/L as $CaCO_3$) is known, the $CO_2$ per million gallons can be estimated. First, it must be remembered that a concentration in mg/L can be expressed as parts per million. Therefore,

1 mg $CaCO_3$/L of water = 1 lb $CaCO_3$/1 million lbs water X LBS $CaCO_3$/1 million lbs water × 0.44 lbs $CO_2$/lb $CaCO_3$ × 8.34 lb water/gal water = Y lbs $CO_2$/MG water

b) equivalent basis

1 meq $CO_2$/1 meq $CaCO_3$ × 22 mg $CO_2$/meq $CO_2$ × meq $CaCO_3$/50 mg $CaCO_3$ = 0.44 mg $CO_2$/mg $CaCO_3$

**2. Carbonate alkalinity to bicarbonate (using chemical reaction 2a)**

a) molar basis

1 m mole $CO_2$/1 m mole $CaCO_3$ × 44 mg $CO_2$/m mole $CO_2$ × m mole $CaCO_3$/100 mg $CaCO_3$ = 0.44 mg $CO_2$/mg $CaCO_3$

b) equivalent basis

1 meq $CO_2$/1 meq $CaCO_3$ × 22 mg $CO_2$/meq $CO_2$ × meq $CaCO_3$/50 mg $CaCO_3$ = 0.44 mg $CO_2$/mg $CaCO_3$

**3. Caustic alkalinity to bicarbonate (using chemical reactions 1a and 2a)**

a) molar basis

2 m mole $CO_2$/1 m mole $CaCO_3$ × 44 mg $CO_2$/m mole $CO_2$ × m mole $CaCO_3$/100 mg $CaCO_3$ = 0.88 mg $CO_2$/mg $CaCO_3$

b) equivalent basis

2 meq $CO_2$*/1 meq $CaCO_3$ × 22 mg $CO_2$/meq $CaCO_3$ × meq $CaCO_3$/50 mg $CaCO_3$ = 0.88 mg $CO_2$/mg $CaCO_3$

The use of these types of calculations for estimating the $CO_2$ requirements for recarbonation is illustrated in the three examples provided in Chapter X. These examples should be studied to better understand where $CO_2$ is added in various water plants, and what happens to the alkalinity during recarbonation.

## Types of Recarbonation Facilities

There are several types of facilities available for recarbonating lime softened water. The means of introducing $CO_2$ to water has undergone considerable change over the past few decades. Since older, existing water treatment plants still utilize facilities that are no longer being

*NOTE: Since two separate chemical reactions (1a and 2a) are involved, a total of 2 meq of $CO_2$ is required for each meq of alkalinity.

designed today, both these facilities and the facilities that are in common use currently are discussed.

## *1. Older Types of Facilities*

The principle used in many of the original facilities was to generate a gas containing carbon dioxide by burning fossil fuels (e.g., methane, oil and coke) in the presence of excess air. This gas was then treated and blown under pressure to the point of application. The method of generating the gas and the degree to which it was processed prior to being introduced to the water varied depending in part on the size of the water treatment plant.

Perhaps the simplest type consists of a tall tank provided with a gas burner, another similar tank for cooling the gases, a compressor or blower, and a diffuser. A plant of this kind using hot water heating tanks is illustrated in Fig. 13-1. Recarbonation equipment of this type has been used when the quantity of water to be treated was less than 10,000 gallons per hour.

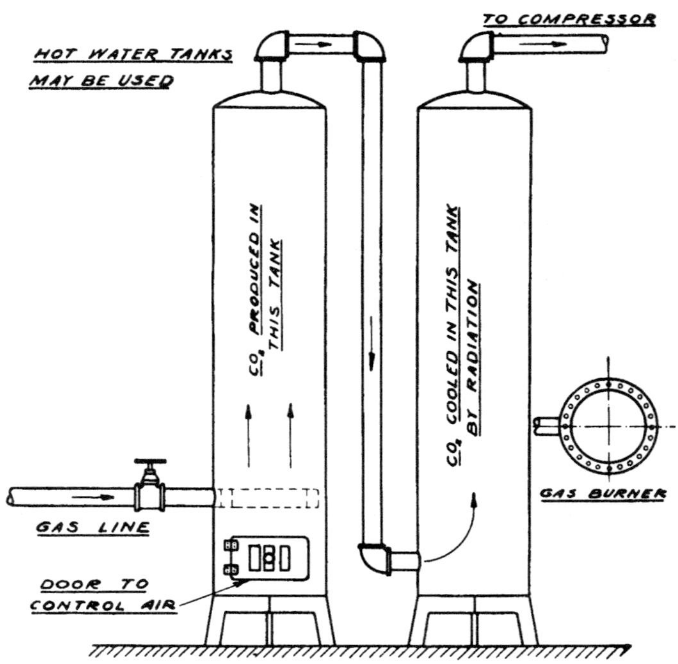


FIGURE 13-1. GAS BURNER CARBONATION PLANT.

At larger water softening plants a gas producer type plant was often used to advantage. This type of plant has six principal parts: a gas producer, a gas burner, a return tubular boiler, a steam-driven air compressor, a combination scrubber and drier, and a diffuser. In operation, coke is burned in a gas producer, which is a closed furnace with controlled air supplied by a blower. The products of combustion pass from the gas producer to a gas burner where they are mixed with air and burned to complete combustion under a boiler. The boiler steam is used to drive the air compressor and blower in order to make the process cyclic. The air compressor takes its suction from the boiler stack of the producer gas-burning boiler, through a scrubber and drier, and forces the washed gases through diffusers into the water to be carbonated. Making producer gas from coke and burning it under a boiler is perhaps the most economical method of producing carbon dioxide gas, because the steam generated is readily available and it is possible to produce a gas containing as much as 18 to 19 percent carbon dioxide, thus requiring an air compressor of much less capacity than if coke were burned in an open furnace. This type of plant is shown in Fig. 13-2.

A third method used in older type water plants is the pressure type of carbon dioxide generator. The proportion of fuel to air is carefully regulated so that complete combustion takes place. The gas is forced to the diffusers by the pressure in the combustion chamber. A sketch of a pressure carbon dioxide generator is shown in Fig. 13-3.

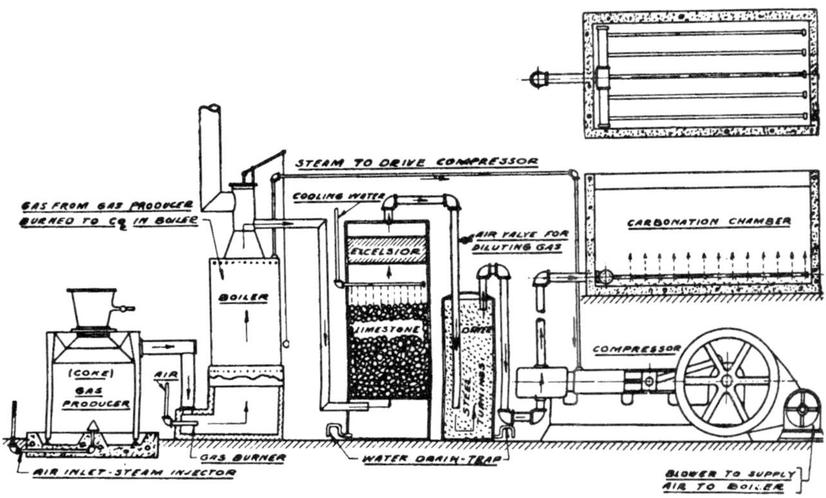


FIGURE 13-2. GAS PRODUCER CARBONATION PLANT.

The equipment requirements of these original types of facilities are extensive. Considerable maintenance is needed to operate these systems. Also, the presence of sulfur and phenolic materials in some of the fossil fuels used to generate the gas containing $CO_2$ periodically causes taste and odor problems in the finished water. Therefore, these types of carbonation facilities are not being designed today for use in water plants.

## *2. Current Types of Facilities*

Two types of recarbonation facilities are predominant in water treatment plants today:

1. underwater burners, and
2. liquid $CO_2$ systems.

Natural gas (methane) is the fuel used to generate $CO_2$ in underwater burners since it does not contain impurities that could contaminate the water. The natural gas and air are mixed and blown to an underwater burning apparatus at the bottom of a recarbonation basin. The gas mixture is ignited in the apparatus and dispersed into the water. In some cases the gas mixture is burned in a unit above ground and blown into an underwater diffuser. This helps mitigate potential problems with the accumulation of carbon monoxide that is produced when the gas mixture is burned.

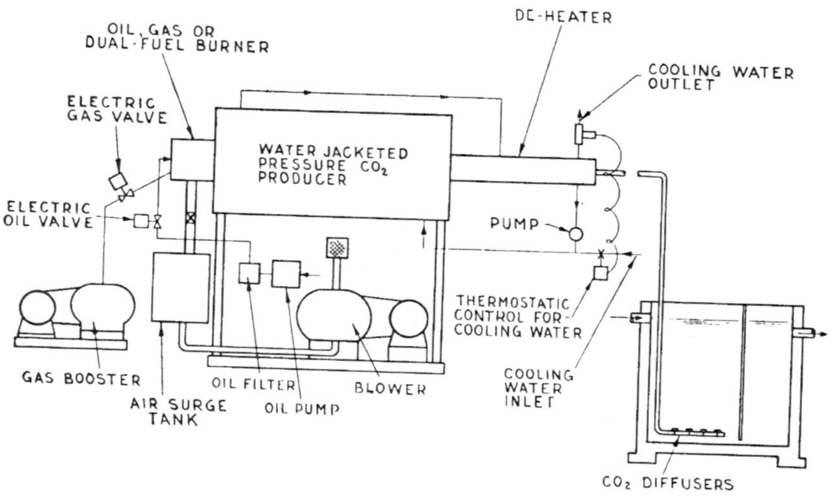


13-3. CARBONATION LAYOUT-PRESSURE $CO_2$ GENERATOR.

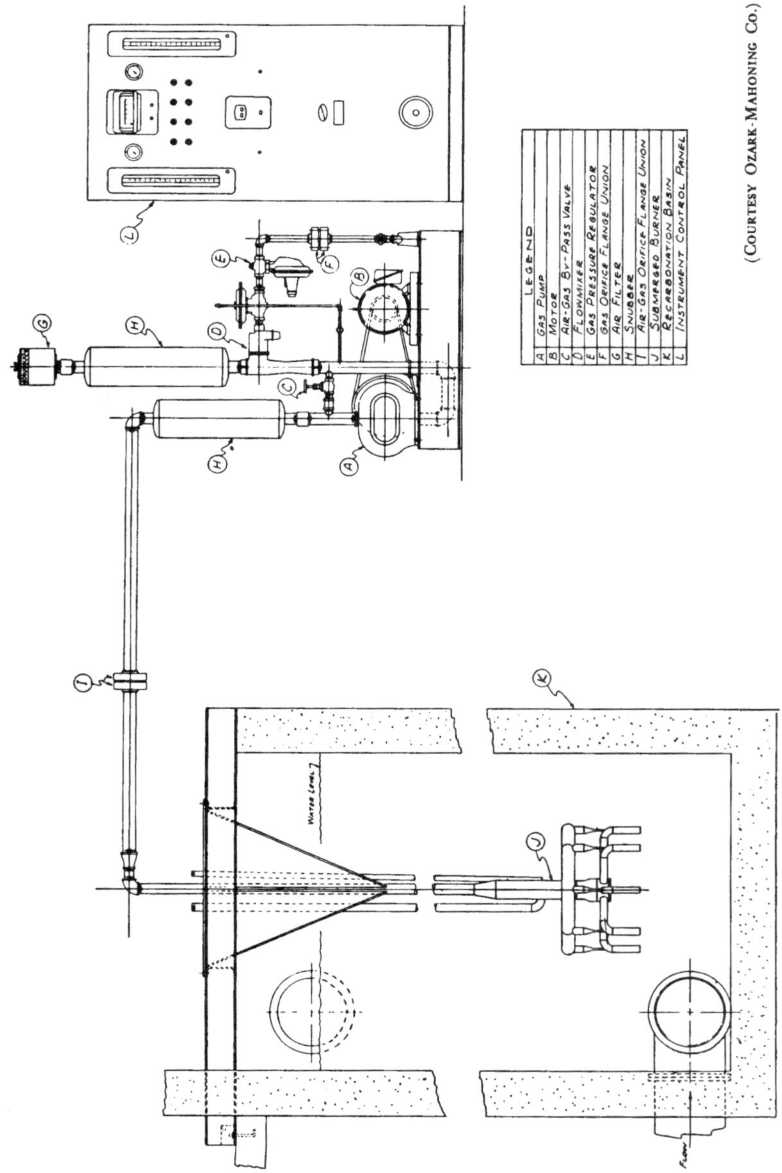

| | LEGEND |
|---|---|
| A | GAS PUMP |
| B | MOTOR |
| C | AIR-GAS BY-PASS VALVE |
| D | FLOWMIXER |
| E | GAS PRESSURE REGULATOR |
| F | GAS ORIFICE FLANGE UNION |
| G | AIR FILTER |
| H | SNUBBER |
| I | AIR-GAS ORIFICE FLANGE UNION |
| J | SUBMERGED BURNER |
| K | RECARBONATION BASIN |
| L | INSTRUMENT CONTROL PANEL |

(COURTESY OZARK-MAHONING CO.)

FIGURE 13-4. GENERATOR USING UNDERWATER BURNER

The quantity of $CO_2$ generated in these units is approximately 0.115 lbs per cubic foot of natural gas burned. The generation of $CO_2$ can be regulated by varying the supply of natural gas to the unit. However, the turndown ratio of typical units is limited 2:1. Therefore, significant variations in $CO_2$ requirements must be met by providing multiple underwater burners. The $CO_2$ transfer efficiency into the water for most of these units is approximately 75 percent.

The use of liquid $CO_2$ in recarbonation eliminates the need of combustion and pressure equipment. Liquid carbon dioxide can be purchased in cylinders. Liquid $CO_2$ systems consist of storage cylinders, refrigeration unit vaporizer, vapor heater, and gas controls and meter. As the carbon dioxide changes from the liquid to the gaseous state, pressure is developed which carries it from the generating cylinder through a heated coil submerged in warm water and also through pressure-regulating valves directly to the diffusers. The amount used can be easily controlled and regulated with a rotameter. Being pure $CO_2$, and not diluted with nitrogen or excess air which are present when the carbon dioxide is produced by burning fuel, the volume of gas to be handled is considerably reduced. Therefore, the carrying and diffuser pipe capacities can be held to a minimum, and the danger of producing carbon monoxide is eliminated.

A wide range of $CO_2$ dosages can be provided easily with a liquid $CO_2$ system. Typical $CO_2$ transfer efficiencies into the water using fine bubble diffusers or a dispersion mixer and gas sparger plate are 80 to 85 percent.

Capital costs for liquid $CO_2$ and underwater burner systems are similar. The cost for liquid $CO_2$ is comparable with the cost of natural gas for generating equivalent amount of $CO_2$ in underwater burners. However, the electrical costs for refrigeration and vapor heating in liquid $CO_2$ systems are generally less than the electrical costs of operating the blower required with underwater burners.

## Recarbonation with Sodium Bicarbonate

Excess caustic alkalinity can be neutralized by sodium bicarbonate in the reaction:

$$Ca(OH)_2 + 2NaHCO_3 \rightarrow CaCO_3 + Na_2CO_3 + 2H_20.$$

This treatment is more expensive than the application of $CO_2$; however, it is useful where recarbonation is needed only at certain times of year. Where lime is utilized to remove the bacterial load of a supply, particularly during late winter and early spring, the amount of lime required may be somewhat greater than is permissible to leave in the finished water. In such cases, the additional chemical cost of bicarbonate treatment may be less than the labor cost of maintaining the $CO_2$ systems in operating condition.

Since one mole of soda ash is formed for every two moles of bicarbonate used, this additional softening reaction reduces soda ash requirements, but requires mixing and sedimentation time following the application of this treatment.

Actual experience has shown that the quantities of bicarbonate required approach the theoretical calculations very closely. For computational purposes one grain per gallon of sodium bicarbonate added to convert $Ca(OH)_2$ to $CaCO_3$ reduces the soda ash requirements by 0.6 grains per gallon.

There are three disadvantages to the use of bicarbonate for this purpose: higher cost than recarbonation; an additional precipitate is formed in the secondary settling basins, increasing the possibility of more $CaCO_3$ colloids being carried to the filters; and in low non-carbonate hardness waters, the bicarbonate dosage required (and resultant soda ash formed) may lower the hardness more than desired.

Chapter XIV

# Ion Exchange (Zeolite) Process of Water Softening

The early types of zeolites were naturally occurring aluminosilicates and were used to extract calcium and magnesium from hard water and replace them with sodium. Later, the zeolites were completely replaced with synthetic resins: beadlike material which exhibits greater exchange speed, longer life, and higher exchange capacity. These new types have important applications, among which are the removal of specific ions such as fluoride, nitrate, and hydrogen (pH). As such, it is incorrect to think of ion exchange solely in terms of water softening. The Safe Drinking Water Act has produced a greater need for these processes, most of which will be discussed later and in other chapters. Because the term "zeolite" is more generally associated with the sodium exchange cycle and especially in the older manuals, the more comprehensive term "ion exchange" is used since it defines the principle involved with all the materials, old and new.

Several hundred different resins are available for the ion exchange process, and the polystyrene are by far the most popular. This Sodium cycle ion-exchanges process is of particular value for bicarbonate waters because hydrogen replaces the alkaline bases to produce $H_2CO_3$, which disassociates into $H_2O$ and $CO_2$. The $CO_2$ can be eliminated by aeration or in a degasifier, and the effluent will be practically the same as distilled water.

With waters containing non-carbonate hardness (sulfates and chlorides), the corresponding acids ($H_2SO_4$ and HCl) are formed, and these are objectionable. They may be (1) neutralized with an alkali, or (2) one portion of water (provided it contains bicarbonates) may be passed through a sodium exchanger (thus producing sodium bicarbonate) and the effluent mixed in the required proportion to neutralize the acidity, or (3) these acids may be removed entirely by a special product called anion exchanger or acid removal.

This material is similar in appearance to the cation exchangers and is used like other zeolites; i.e., by passing water through a bed of the material. It adsorbs $H_2SO_4$ and HCl molecules (but not $CO_2$) and when exhausted it is regenerated with an alkali, usually sodium carbonate. The final product approaches distilled water in quality. The cost of demineralizing is considerably in excess of removing the hardness, especially if chlorides and sulfates are present in the water being treated.

The approximate capacity of these organic exchange products, together with the approximate quantity of salt, acid, or soda ash required for regeneration and the approximate amount of rinse and wash water used in their operation, will be discussed in later paragraphs.

## Siliceous Type Ion Exchange Minerals

Water-softening siliceous type ion exchange minerals resemble ordinary filter sand except that the natural product is green in color. The manufactured or synthetic ion exchanger is more porous than the natural material. The grains usually are larger and almost white in color. Siliceous type ion exchanger contains a sodium base which it will exchange for the calcium, magnesium, iron, and manganese in hard water. It is regenerated or returned to its original condition by backwashing with common salt. The regenerating salt solution is not all used in the revivification process, and much of it, together with the calcium and magnesium extracted from the ion exchanger, runs to waste. After the regenerating treatment, the ion exchanger must be thoroughly washed free of salt or acid and then is ready for service. This process of water treatment is sometimes called the base exchange process. It was known originally as the Permutit system because the application of the process to softening was discovered by a German chemist named Gans who made an artificial ion exchange mineral for this purpose which he called Permutit.

## Softening Capacity of Ion Exchange Minerals

Softening capacities are somewhat less if the water is excessively hard, or if it contains large quantities of sodium salts. The capacity of synthetic ion exchanger is less also if the water contains an excess of magnesium hardness. If the magnesium hardness is in excess of one-half of the calcium hardness, compensation for softening capacity usually is made; similarly, compensation is made when actual hardness exceeds 30 grains per gallon and when sodium salts, expressed in terms of calcium carbonate exceed 10 grains per gallon, or when hardness and sodium -eed 30 grains per gallon. If magnesium hardness exceeds one-half of ·um hardness by 10 grains per gallon, this usually is compensated for as being equivalent to approximately 3 grains additional hardness. For higher hardness and/or higher sodium salt content, the compensated hardness is determined by the following formula:

$$\text{Compensated hardness} = ((9 \times \text{actual hardness}) + \text{sodium salts}) \div 8$$

For example, if the actual hardness is 25 grains and the water contains 10 grains of sodium salts the compensated hardness will be 29.4 grains $((9 \times 25) + 10) \div 8 = 29.4$.

The capacity of ion exchange in terms of $CaCO_3$ removal per cubic foot depends to some extent on the pounds of salt used per cubic foot of ion exchange. The capacity rises with increased amounts of salt per cubic foot of mineral, up to a maximum limit. From that point on, the use of greater amounts of salt no longer increases the softening capacity per cubic foot. More efficient utilization of the salt is obtained when the mineral is worked at low capacities. The most economical use of salt usually lies between 0.3 and 0.5 pounds of salt per thousand grains of hardness removed.

## Polystyrene Divinylbenzene Base Exchangers

This resinous base exchanger material has a capacity of 18,000 to 30,000 grains of hardness per cubic foot of material. Having a very low rate of attrition, the replenishment cost is lower than most other exchanger materials. In a carefully controlled test, the life of a silica ion exchange mineral was found to be about 2 million gallons per cubic foot of exchanger while the test of the polystyrene exchanger had not been completed after 3.8 million gallons per cubic foot had been softened. Capacity of the exchanger depends largely on the amount of salt used per cubic foot of material in the regeneration process. As an example, one product is claimed to have a capacity of 24,000 grains when regenerated at the rate of 0.25 lb. of salt per 1,000 grains of hardness removed. By using 0.47 lb. of salt to remove the same amount of hardness, its capacity is increased to 32,000 grains. While this increases the length of runs by 33 per cent, it increases the salt consumption by 88 per cent. For general practice in municipal plants, it is not economical to use excessive salt but it may be advisable, under emergency conditions, when pumping rates are high, or the water abnormally hard.

The concentration of the salt solution also plays an important part in the exchange capacity of the material. If regenerated with sea water or natural brines, the exchange capacity is not as high as when regeneration is accomplished by means of concentrated salt solution.

## Ion Exchange Water Softening Units

An ion exchange water softening unit resembles an ordinary mechanical sand filter with ion exchange mineral substituted for sand. This filter may be the pressure type device or the open gravity type similar to those used at water purification plants. The pressure softener consists of a closed steel cylinder which may be of the vertical or horizontal type. If tanks are to be shipped assembled, the diameter must be limited to about 10 to 11 feet on account of shipping regulations. There are a number of 14-foot steel welded units in operation. Gravity or open top softeners usually are built of concrete and are rectangular in shape, although some round steel units have been built. The pressure type has one advantage over the open type softener in that it is possible to pump the water to be softened directly from the source of supply through the softener into the distribution system, thus obviating double pumping. Gravity filters have an advantage in that it is possible to observe their condition readily, and to see whether or not there is any channeling of the water or brine through the bed, or any mineral being washed out. Each unit should be provided with:

1. Hard water inlet.
2. Soft water inlet.
3. Wash water inlet and outlet (for downflow).
4. Salt solution inlet.
5. Salt solution and rinse water outlet.
6. Meter or rate of flow controller.
7. Sampling cocks on outlet line of each unit.
8. Underdrainage system.
9. Grade gravel or other suitable support for the mineral.

The inlet or outlet devices may be controlled by hand operated valves. In larger plants it is convenient to have these valves operated hydraulically or electrically, and the control for each unit brought to a central operating table. The principal differences between a standard water filter unit and an ion exchange softener unit are:

1. Sand filters are used to remove from water impurities that are in suspension, and ion exchange softeners soften water by chemical action. In other words, sand is used as a strainer medium whereas ion exchange mineral is used as a chemical contact medium.

2. Ion exchange softeners must be equipped for introducing, removing, and rinsing brine from the unit.
3. The depth of the ion exchange mineral in a softener ordinarily is greater than sand in a filter. Filter sand usually is placed to a depth of from 24 to 30 inches, whereas the depth of ion exchange mineral is a minimum of 30 inches, and can be as much as 12 feet. The empty bed contact time, (EBCT) determines the needed depth, and is usually in the range of 1.5 to 7.5 minutes. The reciprocal of the EBCT, that is 1 divided by the EBCT, determines the exhaustion rate, and is usually 1 to 5 gpm per square foot. The thicker the bed, the less will be the area required for any desired capacity, and the less the area, the lower will be the percentage of wash-water required because washing is based on the rate per square foot per minute.
4. Area of openings in an underdrainage system is greater in sand filters than in ion exchange softeners. In the washing operation of sand filters it is customary to use a vertical rise of from 18 to 36 inches per minute or from 12 to 24 gallons per square foot per minute. To accomplish this, the total area of the holes in the underdrainage system in an ordinary sand filter should be approximately 0.3 percent of the area of the filter. Three-eighth-inch holes on 6-inch centers would give approximately this area. Ion exchange softeners are washed at a much lower velocity, from 6 to 8 gallons per square foot per minute.

Under ordinary conditions a rate of 8 gallons per square foot per minute through synthetic materials will give a bed expansion of approximately 50 percent, and a 6-gallon rate through green sand, an expansion of 25 percent. These must not be taken as exact figures because the percentage expansion is influenced by two factors; namely, size of particles and temperature of wash-water. The finer the particles, the greater will be the expansion, and cold water, having a greater viscosity than warm water, has a greater lifting power. Synthetic units should have a freeboard; i.e., distance from surface of mineral to outlet or overflow of wash-water amounting to approximately 50 percent of depth of mineral. Downflow green sand units should have a freeboard of approximately 35 percent, but upflow units should have a freeboard of 50 percent. A lower percentage of wash-water is required with synthetic materials than with greensand, because the area of the bed of synthetic mineral, for a given capacity, is much less than the area required for greensand, and consequently reduces the water required for washing the materials free from brines used in regeneration.

It is customary to make the total area of the holes in the under drainage system from 0.16 to 0.18 percent of the area of the softener. In other words, 3/16-inch holes on 6-inch centers will give about the right area. The loss of head when passing water up through an ion exchange bed at a rate of 8 gallons per square foot per minute, with 3/16 inch holes in laterals, is about 8 feet.

## Upflow versus Downflow Channeling

Many claims have been made by those interested in promoting the sale of ion exchange softeners regarding the merits of upflow vs. downflow operation. As a matter of fact, either upflow or downflow will be satisfactory if the unit is properly designed and the proper flow rates are maintained. In either upflow or downflow units, provisions should be made to backwash at a sufficient rate to suspend the ion exchange bed and effect a complete rearrangement of the particles.

It is customary to operate greensand downflow units at a rate up to 6 gallons per square foot per minute, and to operate greensand upflow units at rates from 4 to 8 gallons per square foot per minute. It is customary to operate upflow synthetic ion exchange units up to 6 gallons per square foot per minute. With this last rate no backflushing may be required, but is thought advisable occasionally to increase the rate for a few minutes to 8 gallons per square foot per minute.

## Salt Storage and Brine Measuring Tank

The salt storage bin at large municipal softening plants usually is a concrete bin with two compartments. Each compartment generally is built to hold at least one carload of salt. Salt weighs approximately 70 pounds per cubic foot, therefore to hold 50 tons the bin must have a capacity of at least 1500 cubic feet. The bottom of the bin usually is built hopper shape. The lower part of the hopper forms a reservoir or collecting trough over which is placed a screen or perforated pad (preferably made of non-corroding material) covered with about one foot of graded gravel: first layer, 3 inches of gravel, 1 1/2 inches in size; second layer, 5 inches of gravel, 1/4 to 1/2 inches in size; and third layer, 4 inches of gravel, 1/8 to 1/4 inches in size. The salt is shipped in box cars in bulk and is shoveled from the cars into the bin. It is customary to place the bins in the open and as near the railroad track as possible, therefore they usually are built so that the top is slightly above the ground level. They may be either open or covered. After the salt has been placed in the bin it is covered with water. The water dissolves the salt and forms a saturated brine solution which passes throughout the gravel into the collecting reservoir. In a single compartment bin this reservoir is built sufficiently large to hold saturated brine to regenerate one softening unit. In a double compartment bin the saturated brine generally is carried to a brine well located between the two compartments. From either the brine well or the collecting reservoir the brine flows or is pumped to the brine measuring tank, from which it is applied to the softening unit.

## Regenerating the Softening Unit

Saturated brine solution at 60 degrees F. contains theoretically 26 percent salt. It is customary to ignore solubility as affected by temperature and to assume an average solubility of 25 percent. This solution has a specific gravity 1.19228. One gallon of saturated brine solution thus contains practically 2.479 pounds of salt. Before applying the brine to the softening unit it should be diluted. The brine as diluted varies from about 5 to 10 percent solution. In determining the salt charge it is necessary to know the number of cubic feet of ion exchange material in the unit, its exchange capacity, and to assume the number of pounds of salt that are to be used per 1000 grains of hardness removed. For example, let it be assumed that the unit contains 1000 cubic feet of ion exchange material, and that it has a capacity in terms of $CaCO_3$ removable, of 3000 grains per cubic foot. The unit will, therefore, have a softening capacity of 3,000,000 grains. If it is assumed that 0.5 of a pound of salt is required per 1000 grains of hardness removed, there will be required 3000 × 0.5 or 1500 pounds of salt to regenerate the unit. Each gallon of saturated brine contains 2.479 pounds of salt, therefore approximately 605 gallons of brine solution will be required to regenerate the unit. This 605 gallons of brine, to produce a 5 percent solution (1.03624 specific gravity), must be diluted to 3480 gallons.

The first step in the regeneration process, unless high rate upflow is used in softening, is to backwash the unit to loosen up the mineral, to effect a rearrangement of the particles, and to remove any foreign material filtered or precipitated from the water.

The diluted brine solution is introduced into the exhausted unit, either through the underdrainage system and up through the bed, or through a perforated manifold over the surface of the mineral. The brine, if introduced at the top of the mineral, is filtered down into the sand. With synthetic ion exchange materials it is customary to hold the brine in the bed for a period ranging from 5 to 15 minutes. However, the brine usually is passed through greensand slowly and continuously.

If upflow regeneration is used, the process is practically the same except the brine is forced up into the bed and discharged from the top. After salting, the unit is put back in service but the water from it is wasted until the effluent shows a hardness of about 1 grain per gallon as determined by the standard hardness test.

## Automatic Regeneration.

Some municipal and many industrial plants have been built and equipped so that each unit is regenerated automatically after each two or three hours of operation. With a plant of this type very little supervision is required and, as the regeneration periods can be made on short frequency intervals, large quantities of water can be softened with relatively small units, thus reducing the construction cost of the plant.

## Specifications for Salt.

For the regeneration of ion exchange water softeners, using saturators, salt should conform to the following specifications:

*Kind:* Rock salt or evaporated salt.

*Color:* White to grayish white.

*Composition:*

1. Not less than 98 percent sodium chloride with a minimum of calcium and magnesium salts.
2. Phenolphthalein alkalinity zero.
3. No grease, fat or oil content.

*Fineness:* Between 8-mesh and 50-mesh, but evaporated salt may have 15 percent through a 60-mesh screen.

Solubility. The salt should dissolve rapidly without packing and form a clear solution.

*Comparative Solubility and Packing Tendency:* Not over 15 minutes in Standard Float Experiment as shown on next page.

*Standard Float Experiment:*

Place a 20-grain sample of salt in the cup formed by covering with cheesecloth the bottom of a cork ring float 3/4-inch thick, 2-3/4 inches outside diameter, and 2 inches inside diameter. Float the cork and sample in a beaker containing 200 ml. of water. Determine the time for complete solution of the salt. Coarse salt usually requires less than 15 minutes to dissolve and fine salt more than 20 minutes.

Other types of regeneration. Only evaporated salt should be used in the salt pots and direct application of salt.

*Ion Exchange Materials Used as Filtering Medium.*

Although it is not regarded as good practice to operate an ion exchange softener as a filter, nevertheless, downflow softeners will remove any slight amount of suspended matter that may be present in the water, or precipitated from it, and possibly without doing any damage to the ion exchange material.

## Iron and Manganese Removal

Many water supplies contain iron which usually is in a soluble ferrous condition, and if the water is pumped directly from the well through the softener without the iron becoming oxidized, it is removed by base exchange. If, however, any of the iron is oxidized to insoluble ferric iron, it will not be removed by base exchange, but must be filtered. It can be filtered by a downflow softener, whereas it would not be removed by upflow. Iron in solution is not oxidized, nor is it apt to be oxidized by passing through a pressure softener or through an upflow open type unit. There is a possibility of some oxidation if softened through an open type downflow softener where the water necessarily is exposed to the atmosphere.

Iron and manganese can be removed from water by applying sufficient potassium permanganate several minutes ahead of the ion exchanger. This process creates higher insoluble states of manganese on which the naturally occurring iron and manganese are oxidized. This is known as the manganese zeolite process, and it differs from ion exchange softening. It is an iron and manganese removal process, and regeneration occurs with backwash of permanganate rather than salt.

PICTURE 14-1. AERATORS, FORT DODGE, IOWA

### *Iron Fouling**

Difficulties experienced with high capacity resinous type ion exchange softeners used on iron-bearing waters indicate that deposition of precipitated iron apparently takes place on and in the particle of exchange material. This results in a loss of exchange capacity that can seriously affect the usefulness of the softener. Reasonably satisfactory relief is offered by chemical removal of the precipitated iron, which treatment, if applied properly and in the necessary frequency, can maintain the unit at or near design capacity. Sodium metabisulfite has been found useful for this treatment, or one may refer to patented products such as "Zeotone", which is marketed by Calgon.

---

* Journal, A.W.W.A., Vol. 47, p. 253, Feb., 1955.

A procedure successfully used to remove iron fouling is:

1. Backwash as usual.
2. Add brine as usual.
3. Dissolve 1 to 2 lbs. sodium metabisulfite for each cu. ft. of ion exchange material in the unit, pass into the unit via the brine injection line.
4. Let stand 1 hour.
5. Backwash until water is clear (may take several hours).
6. Repeat step 2 (adding usual amount of brine).
7. Rinse as usual.
8. Resume softening cycle.

## Control of Hardness

In operating an ion exchange softener, zero hardness water readily may be produced. The water softened in municipalities, however, is used for all purposes and it seems inadvisable to soften an entire supply to zero hardness. The cost of operation can be reduced considerably if the hardness is blended to 50 to 70 mg/L; in other words, from 3 to 4 grains per gallon. In addition, it is considered necessary to leave some hardness; that is, calcium carbonate, in the water so as to form a calcium carbonate coating on the piping to protect it from corrosion.

The following methods for controlling hardness have been used:

1. Mixing zero hardness water with unsoftened water in the correct proportions to obtain water of the desire degrees of hardness. This is done successfully at some ion exchange softening plants but the procedure is not advisable when the water being softened contains a large quantity of iron, because the mixture will contain iron in direct proportion to the quantity of raw water added. Public water supplies should contain not more than 0.3 mg/l of iron as Fe. Unless the iron-bearing raw water is extremely hard, thus requiring only a small portion to be mixed with the zero water, the mixture of a large percentage of raw water with the softened water would be unsatisfactory on account of the high iron content.

(COURTESY THE PERMUTIT CO.)

PICTURE 14-2. AUTOMATIC ION EXCHANGE WATER SOFTENER

2. Aeration and filtration through sand of that portion required to be mixed with zero hardness water.
3. Addition of enough lime to precipitate iron and allowing it to settle before mixing with the zero hardness softened water. Long settling periods would be required by this treatment.
4. Addition of enough lime to precipitate iron, and filtration through sand filters before mixing with the zero hardness softened water.
5. In one Ohio municipality, where the hard well water supply contains both carbonate and non-carbonate hardness, and iron in solution, all of the iron and the greater part of the carbonate hardness is removed from its entire supply by lime softening. This is followed by recarbonation and sand filtration, after which a portion of this water is then made in the proper proportion to give the desired hardness. This process is economical because carbonates are removed more cheaply by lime than by ion exchange material. Thus each kind of hardness is removed by the least expensive known process. This practice is giving satisfactory results.

(COURTESY THE PERMUTIT CO.)

PITURE 14-3. VERTICAL PRESSURE FILTER IN A RAYON PLANT

6. Ion exchange material will remove iron from water for a time, even after it has become exhausted for removing hardness. In an attempt to take advantage of this, one plant in Ohio, having a hard well supply containing iron in solution, was equipped with four ion exchange softening units. The operation of these units was arranged so that three were softening to zero hardness and the fourth not softening but removing iron. All four discharged their effluent into the same supply reservoir. The difficulty with this scheme is that ion exchange material does not have a sharp exhaustion point. In other words, the hardness of the effluent from an ion exchange softener does not change instantly from zero to that of the raw water supply, and therefore, it was found difficult to maintain uniform composition.
7. Filtration of make-up water through ion exchange material which removes iron but not hardness. These units are regenerated with permanganate as previously mentioned.

## Adjustment of Calcium Carbonate Content and pH Value for Corrosion Control.

Ion exchange zero softened water, to be non corrosive or incapable of dissolving iron from the distribution system, must have its pH value and calcium carbonate content adjusted. If the pH is high, very little calcium carbonate is required for saturation, while on the other hand, if the pH is low, much is required. A curve by Baylis shows that the approximate solubility equilibrium of calcium carbonate varies from about 15 mg/l to over 350 mg/l, between the pH values 6.8 and 9.3. At a pH value of 9.0, for instance, water is saturated with calcium carbonate when it contains only about 20 mg/l, whereas 350 mg/l is held in solution if the pH is 6.8.

Considerable under saturation with calcium carbonate results in corrosion and solution of iron, whereas super saturation builds up excessive scale.The pH value may be raised by aeration to eliminate carbon dioxide, or by the addition of an alkaline material (alkaline water, soda ash, lime, or sodium silicate). It may be lowered by the addition of acids. In lime-soda softening it almost always is necessary to lower the pH value by the addition of carbon dioxide to prevent supersaturation and the formation of too much scale. The alkalinity of stabilized water, or water in chemical balance, that is suitable to be pumped into a distribution system should not materially increase or decrease when treated with powdered calcium carbonate. If it increases, the water is apt to be corrosive; if it decreases, it will be scale-forming.

Well water devoid of oxygen and containing carbon dioxide, if softened to zero hardness by ion exchange and pumped into an iron distribution system, will dissolve oxides of iron from the mains in such quantities as to be entirely unsatisfactory for domestic use. The water should contain a trace of oxygen.

## Study of Water Supply

The water to be softened should be studied before any attempt is made to design a municipal ion exchange softening plant. If the hardness is to be controlled by by-passing raw water, the amount required to produce the hardness desired must be determined. If this amount introduces too much iron, equipment should be installed to remove the iron from the by- passed portion. Chemical balance tests should be made on mixtures of raw and softened water to make sure the mixed water will not be corrosive. In all instances, it is advisable to provide equipment so that alkaline material can be added to adjust properly the pH reaction of the water.

No attempt should be made to soften turbid or polluted water (by the exchange process), or water containing oil or hydrogen sulfide, as these substances tend to deposit on the mineral and clog the pores. Excessively high alkalinity and high hardness water softened by ion exchange may not have an agreeable taste.

## Choice between Lime-Soda Ash and Ion Exchange Softening

The choice between lime, or lime-soda ash, and ion exchange for municipal water softening is one that requires careful consideration in each case. No intelligent choice can be made between he two processes unless the following information is available:

1. Is there land available for the disposal of lime-soda softening sludge? If not, then what disposition is possible for the waste salt brine, barring, of course, locations near the ocean, where disposal can safely be made into seawater or brackish water?
2. Is the water to be softened always clear or is it sometimes muddy?
3. Is the water safe bacterially?
4. What is the analysis of the water? Is it excessively hard? Does it contain large quantities of sodium salts? Is the proportion of magnesium hardness excessive? What percentage of the hardness is carbonate and what percentage non-carbonate? Does it contain silica, much free carbon dioxide, iron, manganese, or hydrogen sulfide?
5. Cost of salt, lime, and soda ash.
6. Is the supply of available water scarce or abundant?

Before EPA restraints in the 1970's, unlike lime sludges, there was no problem in disposing of salt brine wastes. The normal method of disposal now is into the sanitary sewer through a metering arrangement. On the coast of oceans and other salt water bodies, direct discharge is a possibility. In arid regions, properly lined evaporating ponds are in use.

There are also developing beneficial uses for waste lime sludge in agricultural liming and acid neutralization in lieu of lagooning.

If the water to be softened is turbid, a lime-soda ash plant probably can be built much more economically than a combines filtration and ion exchange softening plant.

If a water supply is not safe bacterially, the bactericidal value of lime treatment renders the excess lime softening process suitable for correcting this condition.

Many water supplies are from wells dangerously located with reference to sewers or other sources of contamination. The casings may be very rusty and leaky. The lime softening process is particularly suitable for such wells.

The cost of softening depends on the character of the hardness. Bicarbonate of calcium, for example, is removed more economically than bicarbonate of magnesium, and the bicarbonates are removed with less expense than the sulfates, chlorides, and nitrates.

The cost of lime to remove bicarbonate hardness, under normal conditions, is about one-half the cost of salt when ion exchange is used. On the other hand, the cost of salt to remove non-carbonate hardness, under normal conditions, is about 20 percent less than the cost of soda ash.

## Ion Exchange Household Soft Water Service

Most cities in moderate to hard water areas probably have ion exchange household soft water service. Household ion exchange softening units are replaced periodically; and the exhausted tanks, or ion exchange material, are regenerated at a central regenerating plant.

## Organic Ion Exchange Units.

*Cation Exchanger.*

The capacity rating of these products is dependent upon the analysis of the water to be treated and the operating conditions. Whether the system is operated either on the sodium or hydrogen cycle, the capacity is rated at about 10,000 grains per cubic foot as calcium carbonate (about the same as gel-type synthetic ion exchange materials). Operating on the sodium cycle, the salt consumption is from 0.4 to 0.5 pounds per 1,000 grains of hardness (again this is about the same as for the siliceous ion exchange materials). The acid requirement for water treated with cation medium, when operating on the hydrogen cycle, usually is calculated as follows: The cations are figured to $CaCO_3$ and then 0.45 pounds of 66 Degree Baume or 0.54 pounds of 60 Degree Baume sulfuric acid are required per 1,000 grains.

The acid can be diluted to a ten percent solution in an acid vat and further diluted to about a four percent solution by an injector when being added to the bed. A four to five percent acid strength appears to be about the optimum.

The backwash rate for the cation exchange material is low because of its lightness, and depends upon the freeboard available. An upflow rate of 8 to 10 gallons per minute per square foot is sufficient to loosen the entire bed thoroughly and not result in an excessive bed loss. Usually one hour is allowed for each regeneration of the cation bed. Approximately 37.5 gallons of water per cubic foot of exchange material are required for washing and regenerating.

*Anion Exchanger or Adsorbent.*

The anion exchanger is a porous granular material and is graded between screens having 14 and 48 meshes per square inch.When the anion exchange bed has taken its full capacity for adsorbing acids, it must be regenerated with a caustic solution, usually soda ash. It is sometimes claimed, however, that caustic soda gives better results because with soda ash regeneration the large quantities of carbon dioxide released tend to cause bubbles which channel the bed.

The rating of the anion exchange material is dependent upon the ratio of chlorides to sulfates in the raw water, and upon the operating condition. Usually is has an exchange capacity of around 30,000 grains per cubic foot as calcium carbonate. The soda ash requirement for water treated with the anion medium usually is obtained as follows: The anions are calculated to $CaCO_3$, and then 0.46 pounds of soda ash (98% $Na_2CO_3$) are required per 1,000 grains.

Approximately 75 gallons of water per cubic foot of anion exchange material are required for washing and regenerating.

## Uses of Ion Exchange

The largest application of ion exchange is certainly found in the area of water softening—the deliberate exchange of Calcium and Magnesium ions for Sodium. It is used both on the commercial and residential scale. It is being used as a point of entry (POE) process as well as a point of use (POU) (e.g., hot water heaters, etc.) process. However, it is known that resin beds which are packed with Chloride type resins are used to remove nitrates, chromates, arsenates, and selenates, and the activated alumina resins find a large market in the removal of fluorides.

Activated Alumina exchanges surface hydroxides for fluorides in a process which is essentially adsorption or ligand exchange. These crystals of aluminum oxide are used in sizes of 0.3 to 0.6 mm., and give the added benefit of the ability to remove arsenic. They are usually regenerated with caustic soda (NaOH). It is interesting to note that usual adsorption rates onto granular activated carbon is measured in hours, but adsorption onto the aluminas is accomplished in just a few minutes.

Barium and Radium are removed by ion exchange much more readily than Calcium and Magnesium. Care must be taken with Barium rich waters to regenerate with sufficient NaCl as Barium will accumulate in the bed with insufficient regeneration. Radium is removed at a greater than 95% rate, therefore ion exchange has found new promise as a method for meeting the SDWA limits for that contaminant. In all pilot scale test, Radium never broke through the exhausted bed before the hardness did. That means that expensive testing for Radium can be dispensed with in favor of simple hardness testing.

The *Safe Drinking Water Act* will have a continued impact on the removal of nitrates by ion exchange. Currently, only a few community and non-community supplies use ion exchange for nitrate removal. However, an AWWA survey indicates that exceedance of the 10 mg/l nitrate level was the most common reason for the shutdown of small community water supply wells, and it is assumed that ion exchange will be the only answer for those supplies which choose to stay in the water purification business, the other alternative being the purchase of bulk water from a larger supplier.

Water that is low in total dissolved solids and sulfates is easily subjected to ion exchange for the removal of Arsenic. Preoxidation of the water is necessary, and chlorine may be used for this purpose.

Selenium is another metallic ion which may be removed by ion exchange. While not prevalent in water supplies at the SDWA MCL of 0.01 mg/l, there have been a few instances of its presence. All of these occurrences were in groundwater, and all were naturally rather than man made contaminations.

Finally, chromates in the form of Cr(VI) can be removed efficiently from well water by ion exchange. In fact, in a table published in the Duolite Ion Exchange Manual by the Diamond Shamrock Company, we get an indication that chromate is the most preferred anion by the polystyrene divinylbenzene resins.

Chapter XV

## Treatment of Boiler Feed Water

Large boilers carrying very high steam temperatures and pressures require pure feed water: more pure than can be made in conventional water plants. Boiler feed water used under moderate pressures can be softened and treated so that:

1. It is clear—suspended solids are removed.
2. It is entirely free from non-carbonate hardness.
3. Total hardness does not exceed 35 mg/l.
4. It is practically free from oil.
5. It is practically free from oxygen.
6. It has a low silica content.

Simple coagulation and filtration will remove most suspended solids. Practically all the hardness should be removed from water to be used in extremely high pressure boilers.

Degasifiers are commonly used to remove gas mechanically rather than chemically. Hard water softened to 35 mg/l (cold lime-soda ash process) usually contains from 35 to 100 mg/l caustic alkalinity as sodium hydroxide. If ion exchange softening is used, the water so softened contains sodium bicarbonate. If simply lime softened, the alkalinity is mostly carbonate. The use of these waters will result in considerable priming and foaming (See page 218) unless a regular and adequate blow-off schedule is maintained.

## Boiler Scale

Boiler scale reduces heat transfer and may cause overheating of the boiler metal on which it has formed. This is especially true in high pressure boilers, as is shown in the table below. The density and porosity of scale affect its heat conductivity and the degree to which it causes overheating of the boiler metal. Dense scale causes less decrease in heat utilization and less overheating of the boiler metal than porous scale. During boiler operation the interstices of porous scale are filled with vapor, which results in a lower conductivity.

The following table shows the thickness of scale that can be tolerated without overheating tubes. (The maximum continuous temperature for low carbon steel boiler tubes is about 900°F.)

| *Boiler Pressure Pounds* | *Water Temperature* | *Safe Limit Tube Temperature* | *Scale tolerance in inches* | |
|---|---|---|---|---|
| | | | *Non-Porous* | *Porous* |
| 100 | 328°F | 562°F | 0.117 | ....... |
| 600 | 492°F | 900°F | 0.085 | 0.032 |
| 1800 | 622°F | 900°F | 0.058 | 0.002 |

Water for boiler feed may be softened by the cold process in accordance with the same principles used in softening water for municipalities, or by a hot process, by ion exchange, or by a combination of lime treatment and ion exchange. Many of the cold water softeners now in service for softening boiler feed water were built differently than those for conventional municipal plants. The settling tank in these plants usually is a cylindrical steel tank, sometimes more than 50 feet high,

(COURTESY THE PERMUTIT CO.)

PICTURE 15-1. DOUBLE UNIT COLD LIME PRECIPITATORS.

with a diameter such that the upward flow of water is less than 8 feet per hour. This type plant seldom has baffled or mechanical agitators for prolonged stirring of the chemicals with the water. The softening chemicals are applied to the water at the top of the cylindrical settling tank. Usually the water and chemicals enter a quick-mixing box, which gives the water a whirling motion and discharges it into a down-take tube of sufficient size to retain the mixture of chemicals and water for a period of about 45 to 60 minutes. The velocity and revolving motion of the water as it passes down this reaction tube are sufficient for the formation of a crystalline floc or precipitate.

The flocculated water passes out at the bottom of the down- take and then flows upward into the settling compartment of the tank. Six hours settling, provided the proper chemical treatment has been applied, ordinarily is sufficient to produce a satisfactory boiler feed water. Where settling periods are shorter, filters frequently are used to remove suspended solids. They usually are the pressure type. A typical plant of this kind is shown diagrammatically in Fig. 15-1.

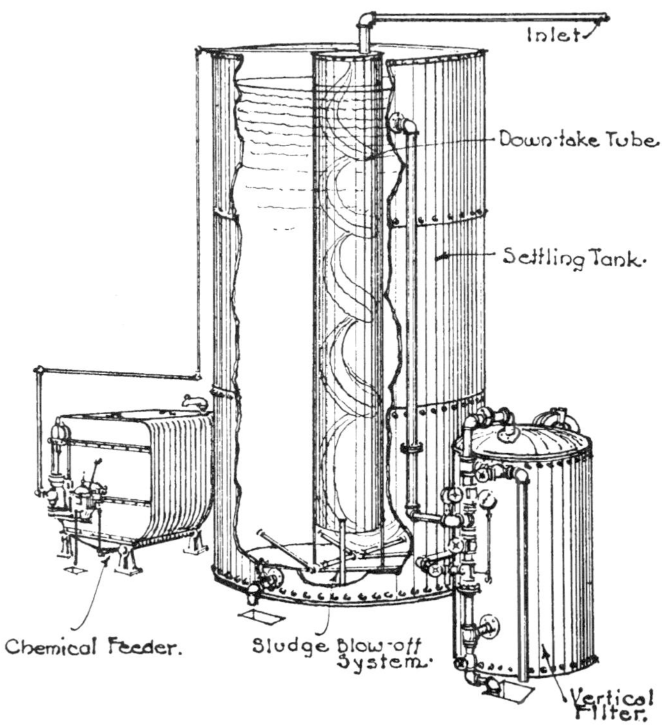


FIGURE 15-1. HOT LIME-SODA PROCESS

Many sludge blanket or up-flow softening plants have been built recently for softening boiler feed water. This type of equipment is recommended by the manufacturer as being especially adaptable when dolomitic lime (high magnesium lime) is used for the removal of silica from the boiler feed water.

To produce water having a hardness of 25 to 35 mg/l by the cold process, the chemical treatment involves the addition of from 35 to 100 mg/l excess lime and sufficient soda ash to remove all non-carbonate hardness, and to convert the excess lime ($Ca(OH)_2$) to sodium hydroxide ($NaOH$). It may be necessary to use a compound of aluminum unless the settling time is long enough for complete crystallization of precipitates. The quantity used rarely exceeds one-half pound per 1,000 gallons. Sodium aluminate usually is preferred as its addition to water does not increase the non-carbonate hardness. However, if the sulfate ration is low in the water being treated, aluminum sulfate should be used.

## Lime-Soda Ash Softening for Boiler Feed Water (Hot Process)

Temperature is one of the most important factors in the lime-soda ash process of water softening. Water used for boiler feed purposes is heated eventually, therefore it is advantageous to heat it for the softening process. Hot lime-soda ash softening is widely practiced. (*See Figs. 15-2 and 15-3)

A hot water continuous softener is essentially the same as a cold softener. It consists of a heater for heating the raw water; a chemical feeding device for proportioning the softening chemicals according to the rate of flow of the raw water; baffles, or some means for mixing the chemicals and raw water; and a settling tank and filter (usually of the pressure type and sometimes equipped with filter coal or some non-silicious filter material instead of filter sand). The mixing and settling compartments, per unit of capacity, are much smaller in the hot process plant than in the cold.

The principal advantages accorded the hot over the cold process are:

1. Free of carbon dioxide.
2. Chemical reactions proceed more rapidly at high temperatures.
3. More complete reactions are obtained.
4. A more crystalline precipitate, which settles more quickly, is produced.
5. Apparatus being smaller, per unit of volume, less housing space is required.

6. Lower hardness water is obtained without the use of excess quantities of lime and soda ash, resulting in lower alkalinities in the softened water.
7. Dissolved oxygen and nitrogen are reduced or eliminated by heating, or by heating and scrubbing with steam.

Lower alkalinities are an advantage because there is less tendency to foaming and priming induced by concentration of sodium alkalinity.

Phosphates sometimes are fed into the already softened, filtered water, or into the boiler with the softened water to prevent scale formation. The phosphates combine with the calcium carbonate hardness not removed in the softening process to form tricalcium phosphate. Further reference to this treatment will be made under the heading "Phosphate Treatment."

## Silica Removal

In the trend towards using extremely high pressure steam in turbines, all silica must be removed. Silica tends to carry over and deposit

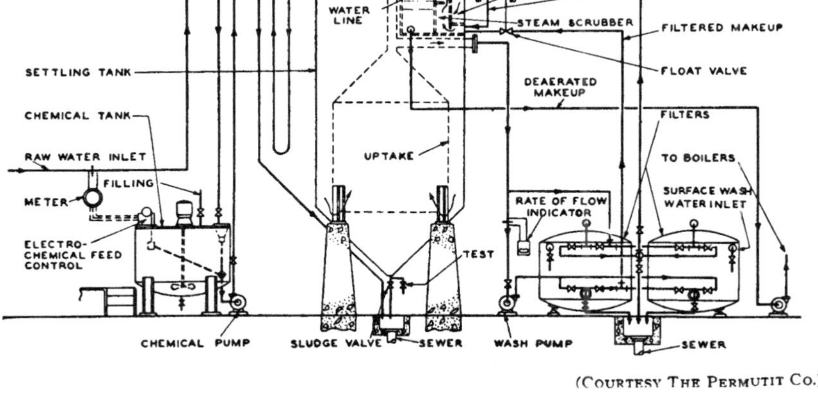


(COURTESY THE PERMUTIT CO.)

FIGURE 15-2. HOT LIME-SODA SOFTENER WITH INTEGRAL DEAERATOR

on the blades of the turbine, and, because of the close tolerances necessary, even a slight build up must be avoided. While distillation removes silica, this is an expensive process. Originally the combination anion-cation exchangers removed practically everything but silica, but newer developments in this field have resulted in almost complete silica removal. In the absence of a condensing turbine, silica removal may not be necessary.

However, since either hot or cold lime process is so widely used in boiler feed water treatment, it has been found advantageous to utilize the absorption power of freshly precipitated magnesium hydroxide (precipitated in the water during treatment). In waters high in magnesium, it is possible that sufficient $Mg(OH)_2$ will be formed to absorb all the silica, but usually additional $Mg(OH)_2$ will be needed.

By using dolomitic lime in the process, sufficient magnesium is introduced to remove the silica. By recirculating the sludge through a magnesia dissolver, better utilization of the magnesium content may be effected.

## Sodium Ion Exchange Softening for Boiler Feed Water

The ion exchange system of softening water for boiler feed purposes is subject to the following objections:

1. Severe pitting and corrosion where it is not possible or practical to remove the oxygen from the water before use.
2. Bicarbonates of calcium and magnesium are converted into a more soluble form, which if allowed to concentrate will cause foaming.
3. Bicarbonates of calcium and magnesium are converted to sodium bicarbonate which, in turn, decomposes in the boiler, if under pressure, into sodium hydroxide and carbon dioxide. Carbon dioxide, if not removed before it enters the boiler, may cause corrosion.
4. Ion exchange units can not be operated with raw water containing turbidity or oil.

The advantages of the ion exchange process are:

1. Virtually all hardness is removed.
2. It is simple to operate.
3. There is no sludge to dispose of.

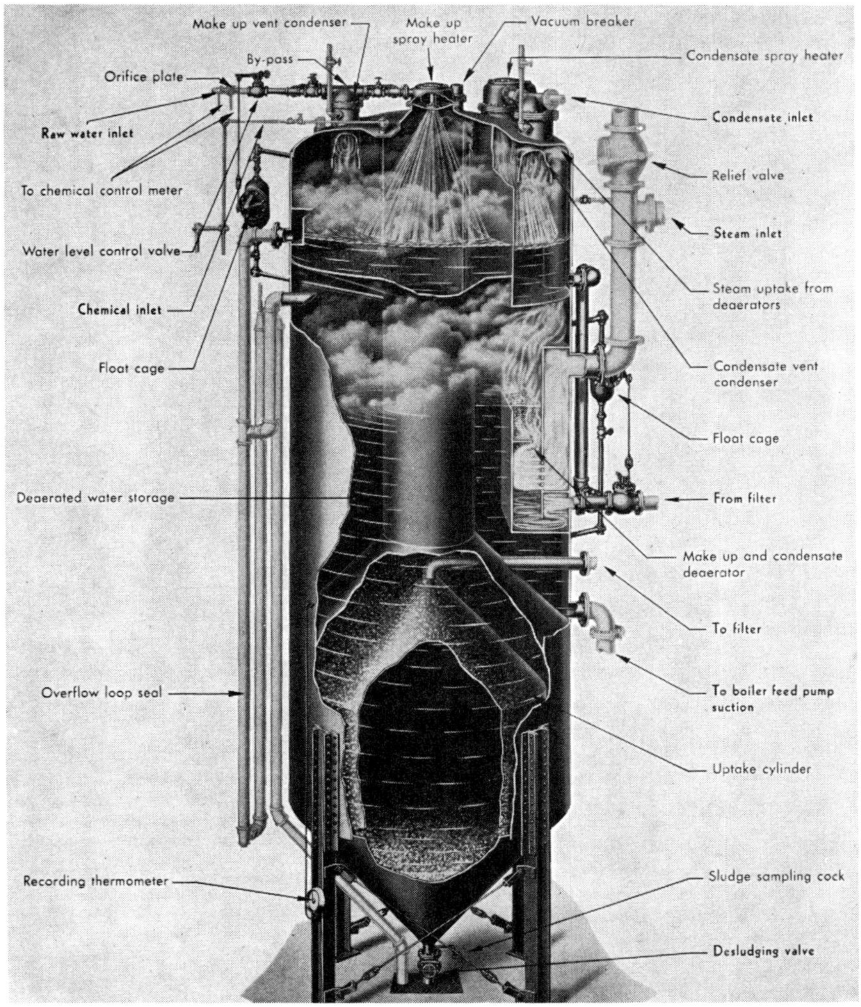


(COURTESY THE PERMUTIT CO.)

FIGURE 15-3. HOT LIME-SODA WATER SOFTENER

## Organic Ion Exchange Materials

Organic ion exchangers (see discussion in preceding chapter) operated on H cycle may be used to remove all metallic cations (Ca, Mg, and Na) but they will be replaced by the corresponding acids $H_2CO_3$, HCl, and $H_2SO_4$. If the acid is all $H_2CO_3$ it can be removed by aeration or in a degasifier. If the acid is HCl or $H_2SO_4$ it may be neutralized by an alkali, by bicarbonate water softened through an ion exchange softener operated on the sodium cycle, or removed by passing through a special anion exchanger or adsorber. By a combination of the cation exchanger and the anion adsorber, it is possible to produce water that closely approaches distilled water in quality.

## Lime-Ion Exchange Softening for Boiler Feed Water

A modified ion exchange system, the lime-ion exchange process, has been developed for softening water with high calcium and magnesium bicarbonate hardness. It possibly reduces the cost of treatment and avoids the difficulties of straight ion exchange softening. The process consists of softening the raw water with lime, settling, carbonating or acidifying, filtering through sand, and then by ion exchange.

The advantages are:

1. The bicarbonates of calcium and magnesium are almost completely removed.
2. Unless investment charges are too high, the cost of treatment is lower because it is cheaper to remove carbonates with lime than with ion exchange. It is also cheaper to remove non-carbonates with ion exchange than with soda ash.
3. Carbon dioxide is eliminated.
4. Turbidity and oil are removed before the water goes to the ion exchange softener.

The disadvantages over straight ion exchange are:

1. Increased investment cost.
2. Operation not as simple.
3. Disposition of sludge.

## Condensate for Boiler Feed

Condensate from steam turbines or evaporators can be returned to boiler feed water heater without any treatment since it contains no oil and is practically free from mineral solids. Condensate containing oil, such as that obtained from reciprocating engines, should have the oil removed before it is returned to the steam system. The oil in the condensate can be removed by coagulation and filtration, or by the use of septum filters. Soda ash and sulfate of alumina are ordinarily used for this purpose in the former process. It is preferable to mix these chemicals with the raw condensate 15 to 20 minutes, then settle for several hours and filter. Fig. 15-4 shows a satisfactory layout for removing oil. The chemical dosages vary with the amount of oil to be removed. The usual charge is 2 to 3 grains of soda ash and 2 to 3 grains of aluminum sulfate per gallon.

Condensate usually has a very low content of dissolved mineral matter due to carryover and condenser leakage, and therefore requires only a very slight dosage of phosphate to react with any hardness that may be present. Boiler blow down, consequently, is slight. It is important that all oxygen and carbon dioxide be eliminated from the water before it enters the boiler. If there is a possibility of any being present it is advisable to add a small quantity of soda ash or caustic soda to insure against corrosion. Some plants operate almost entirely on condensate, the make-up being very small, whereas in others the make-up is very high. Many large plants, in which nearly all the steam generated is condensed for feed water, use evaporators to supply distilled water to make up the unavoidable losses. Such a system transfers the problem of scale from the boiler to the evaporators.

Lime-soda (both hot and cold processes) and ion exchange softened water frequently are used for producing satisfactory make-up.

## Internal Treatment

Internal treatment to control scale formation within boilers is accomplished by coagulation (carbonate), phosphate residual, chelation, coordinated phosphate, or polymer complexation/dispersion. In the coagulation process, sodium carbonate, sodium hydroxide, or combinations of the two are added to the boiler water, supplementing the alkalinity of the unsoftened makeup water. The purpose here is to precipitate calcium carbonate under controlled conditions, thereby preventing calcium deposition in the boiler during latter stages. The newer pretreatment systems are making this method obsolete because of the high cost of frequent blowdown.

The phosphate program involves the creation of an insoluble precipitate through the addition of sodium phosphate. The method is usually employed where internal pressures exceed 250 lb per square inch. This type of treatment is also losing favor because of newer chelating treatments.

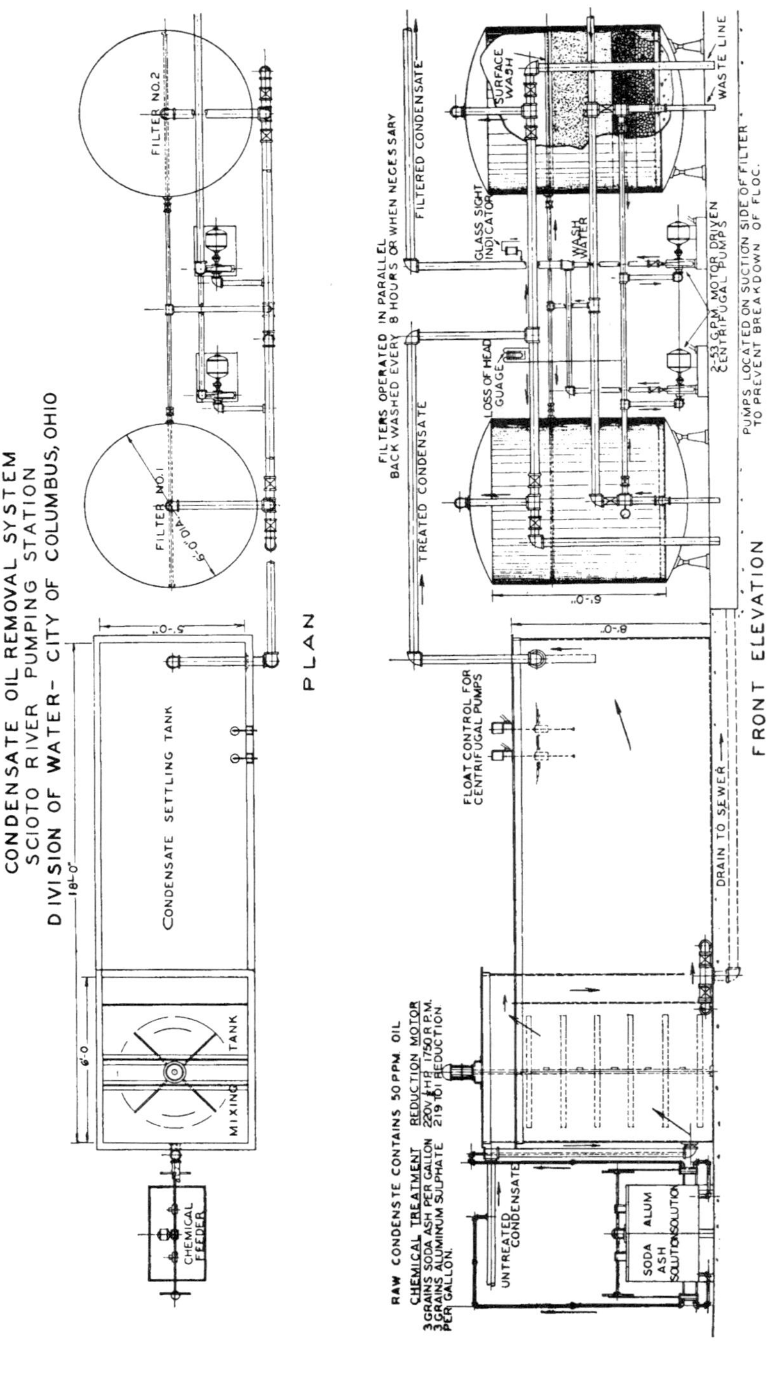


FIGURE 15-4. CONDENSATE OIL REMOVAL SYSTEM

When boilers are operated at below 1500 lb per square inch, and the hardness of the feed water is low, chelants like EDTA and NTA are added to the water to form soluble organic complexes which minimize expensive blowdown. Polymer addition will sometimes enhance the chelation process. The presence of oxygen in these systems necessitates higher chelate feed, and therefore is more expensive.

## Boiler Corrosion

Water containing mineral acids, or salts having an acidic nature such as aluminum or iron sulfate, is very corrosive and destructive in its action when used for boiler feed.

Water from swamps or ponds containing decayed vegetable matter or sewage may not show an acid reaction when cold, but organic acids are formed by the heat in the boiler, resulting in corrosion.

The presence of free carbonic acid is conducive to corrosion but is not as destructive as mineral acids because it is rapidly expelled from the water by boiling. It is said that the presence of magnesium chloride in water sometimes causes corrosion due to the breaking up of this compound at boiler pressure, thus releasing free hydrochloric acid.

If corrosion occurred following the use of acid or swamp water only, the problem of corrosion would be simple because acid is neutralized easily with lime or soda ash. However, many waters with an alkaline reaction are corrosive because they contain oxygen.

All surface water, treated or untreated, contains dissolved oxygen unless it has been removed. The amount of corrosion taking place in a boiler due to oxygen may vary widely. Two feed waters containing the same amount of dissolved oxygen will not necessarily cause the same amount of corrosion. Some of the influencing factors are hydrogen-ion concentration, incrusting solids in the feed water, internal treatment, temperature, and method of operation. Waters high in those salts which upon decomposition produce mineral acids are conducive to corrosion. Especially is this true when nitrates and chlorides are present.

In practice it is found that the maintenance of a pH of 9.5 or higher, and a maximum of 0.01 milliliter per liter of dissolved oxygen, will insure very little corrosion, though not always complete inhibition.

Oxygen may be removed either by chemical treatment or mechanical de-aerators. Common de-aerating chemicals are tannin compounds, ferrous salts, and sulfites. They are especially good for removing the last traces of oxygen. There are various types of de-aerating devices which are described in the literature and in the pamphlets of manufacturing equipment companies.

## Boiler Plate (Caustic) Embrittlement

So-called caustic embrittlement or embrittlement cracking has been a controversial subject of a number of years. Some have advocated that an alkalinity to sodium sulfate ratio be maintained to prevent it. Others have maintained that lignin and other organic materials offered more protection from cracking than did sodium sulfate. Still others have advocated the use of phosphates, tannates, chromates, and nitrates.

More recently so-called embrittlement cracking has been regarded as fundamentally due to poor boiler work and the present tendency to prevent it is to use a controlled amount of sodium nitrate.

## Foaming and Priming

Foaming is the formation of bubbles on the surface of the water. When these bubbles accumulate sufficiently to fill the space above the water, they are carried off into the steam pipes, producing wet steam.

Foaming may be distinguished from priming in that during foaming, the feeding of wet bubbles is continuous; whereas in priming, the action is more or less explosive and usually is due to careless operation of the boiler.

Why do boilers foam or prime? Operating engineers say "too much soda ash" in the water. Water service engineers suggest a number of contributing causes as follows:

1. improper structural design of the boiler, such as limited area of steaming surface or volume of steam space;
2. method of operating boiler, including sudden variation in load, height at which water level is carried, and irregular firing; and
3. composition of the water inside the boiler, including concentration of alkali salts and suspended matter, and occasional presence of organic matter.

Foulk has advanced a theory on foams, the gist of which is: solutions foam when the liquid film formed by approaching liquid surfaces is tough enough to resist their merging. Guided by the principles of this theory it is now possible to make a quantitative measurement of the film forming tendency of a liquid, and also to compare the stability of stability of foams formed in different solutions. Foulk gives the following example: "Two bubbles are caused to issue from orifices placed near each other in the liquid, the film-forming properties of which are being examined. If the liquid contains dissolved solids, the bubbles flatten against each other; that is, they cease their approach when very close because a liquid film is formed between them. If, however, the two bubbles are in a pure liquid, like distilled water, or in a very dilute solution, they do not cease approaching because no film is formed between them and the two bubbles merge together into a larger one as soon as they touch each other." By counting a large number of such contacts of two bubbles in the liquid and noting the number of times the bubbles do not coalesce on touching; that is, the number of times a film is formed between them, a quantitative measure of the film-forming tendency of the liquid is obtained.

Chapter XVI

# Water Conservation

## Need For Conservation

The drought conditions experienced in the decade of the 1980's brought about a change in philosophy for the world's water suppliers. The industry, traditionally known for its ability to supply its customers' demands, found itself becoming concerned with the alteration or control of customer consumption. While most utilities cringe at the thought of water use restrictions, there has occurred a growing acceptance of the need to establish long term conservation techniques and consumer educational programs which will help us to keep pace with the water needs of users. In regions where the cost of source development has increased, and in light of the increasing cost of compliance with water quality standards, conservation has become a standard procedure. It is no longer realistic to believe that the industry can simply construct larger and newer facilities to meet the growing demand for water.

## Conservation Methods

Traditional methods such as meter accuracy placement and testing, leak detection, and use restrictions have been supplemented by conservation methods which have developed in the past few years, such as public involvement and education, the engineering of low volume toilets and showers, the structuring of water rates, and the practice of water reuse. In an article published in 1987 in the *Journal AWWA* entitled *"Water Conservation"*, W. O. Maddaus states that the installation of water meters is generally the first step taken by a municipality to make its water system more efficient and its water rates more equitable. Metering has been demonstrated to be effective in reducing water demands because users pay according to the amount of water they use. He states that data gathered by the Denver, Colorado water department from 1980 to 1982 shows that metered households use about 20% less water annually than unmetered ones. Metering is an expensive operation (installations, reading, and repair or replacement), and so some supplies in this country remain unmetered. However, it is believed that, in spite of high initial cost, metering is proving to be a cost effective conservation method. Meters should be maintained and tested for accuracy periodically. Meters will under register with age, and as such will not encourage the average homeowner to conserve on water usage.

Leak detection and repair will help to save wastage of water in the distribution network. Studies performed on water system losses have shown that 10 to 12% of water savings can be realized through a leak detection program. These programs are expensive, but usually pay for themselves.

All systems should determine their unaccounted for water percentage on a yearly basis, and then make attempts to reduce it as much as possible. Inaccurate meters, water theft, seeping underground storage and evaporation, sewer flushing, firefights and public use are all sources of water loss. Some of these practices can be controlled or modified.

Water rates can have an impact on the willingness of the consumer to conserve on water usage, although water experts generally disagree on how much. Most agree that in order for a rate increase to have a positive effect on conservation, the rates should incorporate the following criteria:

1. they should reflect the real cost of water, present and future
2. they should be related to the amount used, so that the user sees the connection between the amount used and the amount paid
3. rate increases should be accompanied by public education so that all can understand what is involved in obtaining and producing the water supply
4. they should not be flat rates; more demand should produce more cost for added consumption
5. they should not be "declining block rates" for the same reason as above
6. where possible, they should incorporate peak demand pricing: that is, seasonal pricing should be followed. It should cost more for water when it is in higher demand.

It is emphasized here that these measures are intended not for profit-making schemes, but as conservation techniques. Thirty-five to forty percent of the residential water wastage is accounted for in the bathroom. The standard toilet fixture is designed to use 5-7 gallons per flush, but newer units can be purchased which cut that amount by 50 percent or more. A plastic bottle filled with water and placed inside the toilet tank can save 15 to 20 percent of usage. Other toilets on the market are waterless, using chemicals or incineration.

The average shower uses 40 gallons of water. This amount can be reduced by 10% by the installation of a low flow showerhead. Other water savings devices include the low flow faucets with aerators, and pressure-reducing valves. Builders, contractors, and homeowners should be encouraged to use such devices. Some communities have rewritten their building and plumbing codes to require such installations.

Central to all of these conservation measures is the need for more public education. Voluntary commitment on the part of the consumer is essential to conservation, and that commitment can only be had from a consumer who fully understands why it is essential to conserve.

There are many ways to educate the consumer, including newspaper articles, booklets, posters, and bill stuffers. Much of this material is available from AWWA. Membership in this organization is a guarantee that the operator will have up to date, pertinent material at his or her disposal. Operators should commit some time to giving presentations in the schools, as these children are future water conservationists.

Water reuse, the use of water for non-potable purposes, has found its niche in the conservation field. Water in the form of treated sewage effluent is sometimes used for irrigation, watering of golf courses, and in some recreational areas. This procedure is limited by the reluctance of some politicians and residents to trust the system.

## Salt River Project

For many communities in the Western United States, reclamation is a necessity. Reclamation is the "reclaiming" of land for farming by the construction of irrigation projects. The National Reclamation Act of 1902 was passed as a means of providing intelligent use of available water for the growth of the American West. This led to the formation of the Salt River Project (SRP) in Arizona, the largest and most successful reclamation development in the United States.

The Salt River Project patterned its construction of water carrying canals after the Hohokam Indians, the area's ancient settlers who lived in the Salt River Valley from 300 B.C. to about 1400 A.D. As the culture of the Hohokam developed, they began using stone tools to carve out canals which brought life-giving water from the Salt River to their vegetable and cotton fields. In what is considered an ancient engineering marvel, the Hohokam may have constructed more than 250 miles of canals in the Valley to support their population of 3000 inhabitants.

This early successful reclamation project died out as its inhabitants left mysteriously around 1400 A.D. But in the 1860's, the U. S. Military and other inhabitants made their way into the Salt River Valley, and the need for reclamation and water was renewed. One of the settlers, John

JACK SWILLING

W. "Jack" Swilling, found evidence of the ancient Hohokam canals, and realized that they could be cleared and used for their original purpose. He organized the Swilling Irrigation Ditch Company and began using the water to grow crops to sell to the military base. This successful endeavor brought homesteaders to the region as new canals were built to carry water to the freshly cleared farmland. Over the years, 133 miles of main canals and 1,132 miles of laterals and ditches were constructed to serve the area.

Problems arose during dry periods due to lack of supply, and settlers argued over water rights. Those who arrived earlier than others argued that they should have claim to the water first during drought periods, while others argued that all must share equally. Theft and poorly constructed dams ate into the supply, and no method was available to catch the abundant waters from the spring runoff. The Valley citizens, realizing that the farming community was worth saving, pooled their efforts

PICTURE 16-1. NO. 1625 - GATES AT EXIT END OF COTTONWOOD PRESSURE PIPE. SALT RIVER PROJECT—ARIZONA. PHOTOGRAPHED FEB. 26, 1906.

and petitioned the Federal Government for relief. President Theodore Roosevelt, believing that water development in the West was essential to the future prosperity of the nation, helped approve the National Reclamation Act in 1902. Two years later, the SRP was born, and began overseeing operations and supplying water to the farmers. Irrigation fostered reclamation and the area prospered.

The central idea of this project was that each landowner retained rights to a water volume allotment commensurate with the amount of land which he owned. It still operates that way today. As the farmland gradually gave way to urban development, the need for potable water supply grew and the SRP adapted with ingenious methods.

The SRP is and has been the largest deliverer of water in Arizona, one of their customers being the City of Phoenix Waterworks. Over 70 billion gallons of water are delivered each year to the City of Phoenix and other purification plants in Tempe, Scottsdale, Mesa, Glendale, Peoria, Chandler, and Gilbert. In the SRP area, the right to water belongs to the land and cannot be taken to another piece of property. Even when the land is sold it keeps its original right to water. Because residents in the city do not have a way to purify the water, the cities act

PICTURE 16-2. THE HOHOKAM AND THEIR CANALS.

as agents for them and SRP to purify the water and return it to homeowners and businesses.

For example, in 1952, the SRP and the City of Phoenix entered into a contract. The city pays SRP the annual assessment for urban acreage which no longer is irrigated. In turn the Project delivers to Phoenix the water to which this urban acreage is entitled. In this way, the city's potable water needs are met at a reasonable cost and water is conserved that otherwise might be lost in transmission. The city's raw water source is brought to them at the same rate of payment as that of the individual irrigation customer, which saves considerable pumping cost and watershed surveillance.

Chapter XVII

# Water Distribution

In addition to adequate supply and proper treatment, facilities for distributing the water to the consumer require careful attention. It is the responsibility of the water utility to take all possible precaution to assure that the water reaches the consumer in satisfactory quality and sufficient quantity. A distribution system not properly maintained can be a source of problems and complaints, and will undo the good which was accomplished at the plant.

A clean distribution will permit the delivery of a safe quality water to the customer. Routine flushing of the system should be scheduled, and customers should be notified well in advance. A good public relations program is needed for this, as discoloration of the water temporarily assaults the system as it is flushed. It can be emphasized that such flushing is in the best interest of the customer from a public health standpoint. Periodic main flushing will eliminate the nutrients that coliforms need to grow on, and is a must when passivation films such as Zinc Orthophosphate are used for corrosion control. Rules for the flushing of water mains are:

1. Do not flush a large main from a small one.
2. Be careful not to shut down an area or main inadvertently.
3. Never reduce pressure at a fire hydrant to zero when services are connected to isolated mains beyond it.
4. Do not proceed from the last flow at a given fire hydrant until "slugs" of discolored water have been removed from junctures with mains previously flushed.
5. Avoid opening or shutting hydrants rapidly to avoid water hammer.
6. Do not leave an area practically shut down overnight, as this will hamper fire flows.
7. Avoid damage to streets and lawns.
8. Be sure that valves are holding so that flows are coming from expected directions.
9. Mark closed valves on a map, and erase marks when valves are reopened.
10. Watch out for construction projects.

11. Be sure that all water contributing to flows comes from previously cleaned areas.
12. Remember to notify customers.
13. Always match the flow to the pipe size.
14. Keep the length of the main as short as possible.

Studies performed by research microbiologists like Edwin Geldriech of the USEPA and Mark LeChevallier of American Water Works Service Company have shown that the disinfection of biofilms in the distribution system is one of the most important tasks performed. Biofilms are responsible for sporadic or recurrent coliform outbreaks, as well as for the production of corrosion potentials in the system, which lead to main failure. Biofilms can be reduced or inactivated by the maintenance of a chlorine residual, periodic main and dead end flushing, (twice each year), and the addition of corrosion inhibitors such as Zinc Orthophosphate. A water main flushing velocity of at least 2.5 fps should be achieved to avoid biofilm accumulation.

*For example: How many gpm are needed to achieve a 2.5 fps velocity in a 12 inch main?*

$Q = A \times V$, where $Q$ = the flow in $ft^3/sec$, $A$ = area in $ft^2$, and $V$ = velocity in ft/sec.
so $Q = (3.14 \times .5' \times .5') \times 2.5$ ft/sec $= 1.9625$ $ft^3/sec$
and,
$1.9625$ $ft^3/sec \times (7.48$ $gal/ft^3) \times (60$ $sec/min) = 880$ gpm

Adequate records of the size and the exact location of all mains and shutoff valves should be maintained. These records should be kept up to date and should be readily accessible to repair personnel. Notations as to left and right hand valves should be made so the valve can be closed in an emergency without breaking or damaging the valve. Fig. 17-1 is an example of a section of an area map made available to a valve crew that has been sent to do a valve and hydrant study. It is a copy of an overall map which is kept up to date in the office. Any changes made by the crew are noted and dated, and then the changes are made to the original.

Large water systems usually hire consulting engineers to "model" the distribution system on a computer aided design (CAD) process. The computer can then be used to predict flows and pressure under variable conditions, and can be consulted on such factors as pipe size design, efficient storage and booster placement, and even valving arrangements.

Standpipes and reservoirs on the distribution system should be covered and all openings protected with corrosive- resistant screens of fine

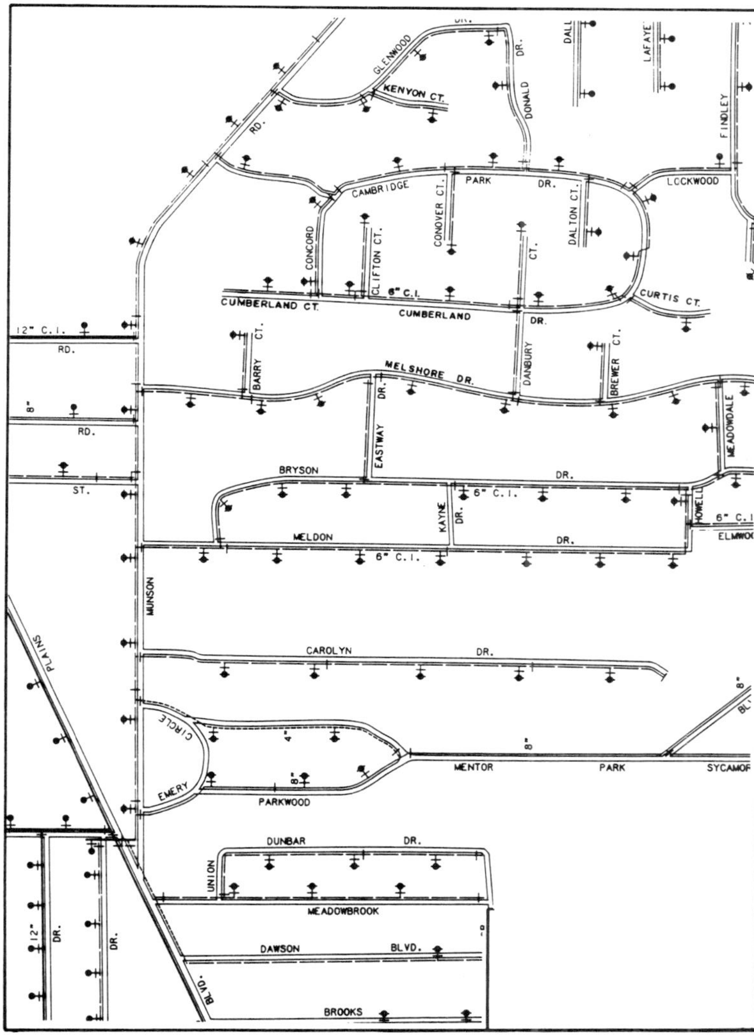


FIGURE 17-1. PART OF GENERAL WATER UTILITY MAP (COURTESY OF B&N ENGINEERS, LTD., COLUMBUS, OHIO.)

mesh to keep out insects. The areas containing these standpipes and reservoirs should be fenced and protected with locked gates. Altitude valves, if used, should be checked to be sure they are functioning properly. Tanks should be inspected frequently for deterioration and should be kept in good repair inside and outside.

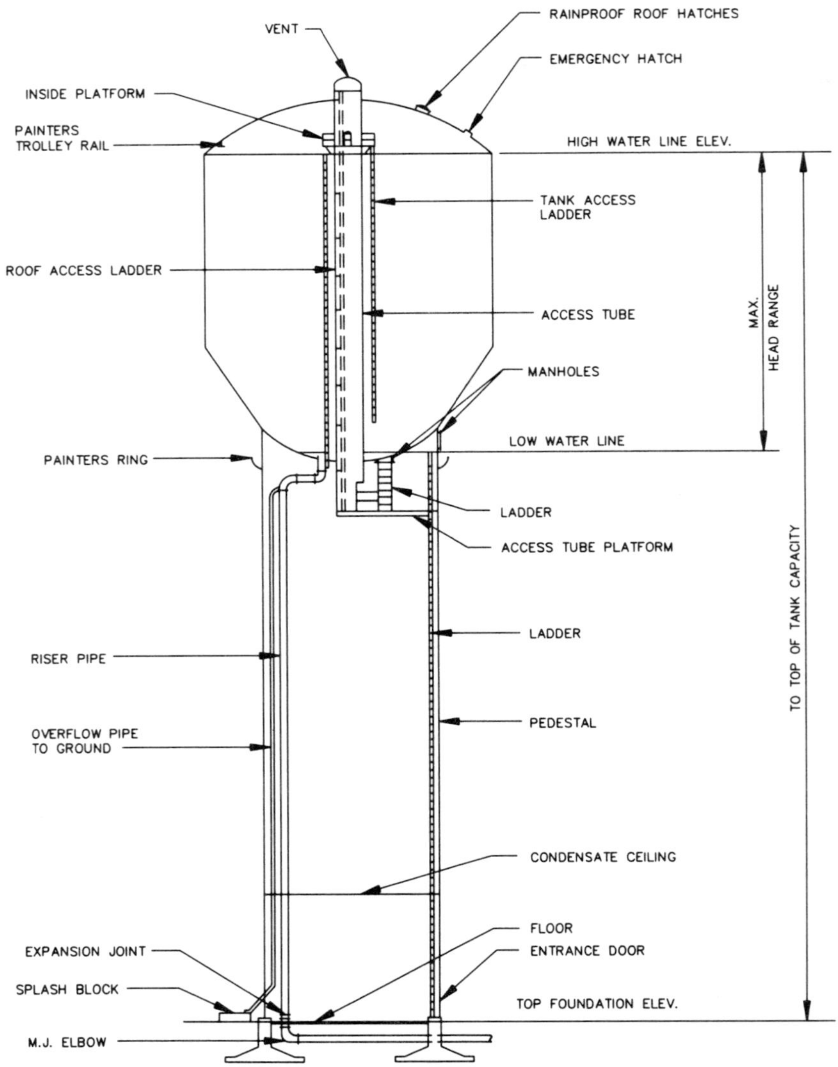


FIGURE 17-2. SCHEMATIC OF TYPICAL ELEVATED WATER STORAGE TANK (COURTESY, BURGESS NIPLE, LTD. COLUMBUS, OHIO).

A schematic of a typical modern 0.5 MG elevated storage tank is depicted in Fig. 17-2.

Pumps and boosters that fill storage tanks should be equipped with adjustable on/off set points or limits so that proper fluctuation or "turnover" of the tank volume can be accomplished. In this way, fresh water with proper chlorine residuals can be maintained. A tank that is continually "topped off" will have stagnant water in it, especially in summer, and will cause taste and odor complaints and possibly bacteria problems. It may be difficult to convince the hard line distribution operator to do this because he has been taught that a full tank means good fire protection, but the public health factor cannot be ignored. If possible, the entire tank volume should be displaced each 24 hours.

All cross-connections must be prohibited. If the consumer has a private water supply, or even reuses some of the city water for other processes, definite precautions should be taken to prevent any possible siphonage that can be induced by opening a large valve on a lower floor, especially when the city line has low pressure, thus allowing a submerged inlet or hose on an upper level to siphon polluted water into the lines. Some old style toilets in hospitals, hotels, and apartments have caused such difficulties and should be replaced.

Dead ends in the system can allow the water to become stagnant, permitting unpalatable or rusty water to be drawn into the system. Where possible, dead ends should be connected with loops so as to permit constant circulation; otherwise, they may require frequent flushing so the need of a fire hydrant or blow off at the extreme end is a necessity.

Private booster pumps in the system should be equipped with low pressure cutoff switches set at 20 p.s.i. so that they will not cause back siphonage. Most states have enacted cross-connection control regulations, and each water supply should adhere to them. The water purveyor should realize that, even though the system can become contaminated from an outside source, all the customers are entitled to protection from such contamination. When a customer purchases water through his meter, he assumes its purity.

All fire hydrants should be installed with adequate drains so that any water drained out to prevent freezing is not permitted to be drawn back into the lines. Even though the water utility may not have the responsibility of installing and maintaining the hydrants, proper surveillance must be made for its own protection.

Good housekeeping practices must be followed in laying new lines. The pipe should be protected before and during installation to prevent undue contamination and should be swabbed out just before laying in place. Steps must be taken to prevent small animals becoming trapped in the line during installation. The new line should be thoroughly flushed and then must be properly sterilized with a high concentration of chlorine (100 mg/l for 24 hours), then thoroughly flushed again until free of excess chlorine. Thorough flushing is dependent on the proper use of adequately sized flush pipes. For example, an 8 inch main cannot be flushed with a one inch blowoff—the proper velocities will not be achieved no matter how long the main is flushed. Refer to the charts in the particular AWWA standard (CLSI) for this information. The line should not be put into service until bacteriological tests show that the water is safe for human consumption.

Similar precautions should be taken in repairing main breaks, replacing valves, or otherwise opening a line. It is easy to pollute the line; therefore, if at all possible, the line should be sterilized after repair or at least very thoroughly flushed before putting back into service. The excavation area should be sprinkled with HTH or other chlorine powder to minimize contamination.

Where treatment in the past has not been adequate, the lines may have their carrying capacity greatly reduced by deposited scale, deposition of corrosion products, or by rough interior due to tubercles and scale. A thorough cleaning program may be needed to restore these lines to their original capacity. If the condition warrants, a relining of the pipe interior may be done. Such work is usually done by organizations who have the equipment and men to do the work properly.

The metering components of the system are the cash register of the business, and should be kept in an accurate working condition. Small

PICTURE 17-1. DISTRIBUTION SYSTEM STORAGE

PICTURE 17-2. HIGH SERVICE PUMP ROOM AT COCOA, FLORIDA PLANT

5/8, 3/4, and 1 inch domestic use meters should be replaced on a regular basis, say each 10 years. These low cost items are simply thrown out or scrapped, having been amortized in that period. This practice is replacing the old meter testing shop which was a part of all systems at one time.

Large meters are considered more permanent items and should be tested on a frequent basis, depending on the amount of water they register. It is not uncommon to see these meters comprise 1 to 5% of the total metered customers in a system, but they may account for as much as 20 to 40% of the consumption. Meter accuracy is affected by meter age, poor installation design, main breaks and fire fighting/hydrant flushing which stirs sediments, aggressive or encrusting water, and customer tampering.

Crenothrix and other forms of higher bacteria may create problems with "red" water due to iron or "black" water due to sulfur. Break-point chlorination may be required to control these problems. In some cases, additional chlorination at points in the distribution system may be required. Use of water pipes for electrical grounding should be discouraged. Serious corrosion can be caused by excessive flow of current through the lines. Institutions such as schools, hospitals, or industries have been known to create serious problems of this type. Also, areas around electrical substations have been subject to severe pipeline damage.

Perhaps the greatest problem in our water distribution systems today is the lack of fully trained and certified operators who work in them. All operators in responsible charge of a water distribution system must be licensed by the certifying authority in that state. A distinction must be made between the repair of a system, and the operations and maintenance of a system. Maintenance is not repair; it is the systematic flushing, valve operation, disinfection, hydraulic operation, computer modeling, backflow device installing and corrosion control practicing in the system. Distribution systems will be as good as the people who operate them.

# Chapter XVIII

# Skeleton Methods of Analysis

## Water Analysis

For detailed methods of analysis, the reader is referred to Standard Methods for the Examination of Water and Wastewater, AWWA 18th edition, 1992. All analyses for reporting purposes must adhere to those found in that text, and as dictated by the certifying authority of the particular state or district in which the water department is operating. These methods are too numerous or complicated to print here, but where methods exist for in plant (non-reportable) analyses, they are enumerated.

*Color*

Color is determined on filtered samples by filling a standard Nessler tube and comparing with standards made of potassium chloroplatinate. Use of Hellige or Taylor instruments for this purpose will be found convenient. Filtration through filter paper may absorb some of the color so it is preferable to use so-called filter cylinders (Berkfield "candles") if available, to remove turbidity with minimum lost of color. *$CO_2$* may be determined by the following methods:

Nomogram—Tillman's formula
Nomogram—Conversion
Acidity—Free Carbon Dioxide

*Free $CO_2$* may be calculated by the Nomograph or titrated by the following procedure.

Pour 100 ml. of the sample into a tall, narrow vessel, preferable a 100 ml. Nessler tube. Add 10 drops of phenolphthalein indicator and titrate rapidly with N/44 sodium hydroxide, stirring gently until a faint but permanent pink color is produced. The free carbon dioxide ($CO_2$) in mg/l is equal to 10 times the number of ml. of N/44 sodium hydroxide used.

*Total Acidity.* Add 4 drops of phenolphthalein indicator to 100 ml. of the sample in a white porcelain casserole. Add N/44 sodium hydroxide from a burette until the solution turns pink. The total acidity in parts per million of calcium carbonate is equal to the number of ml. of N/44 sodium hydroxide used multiplied by 11.4. Acidity in terms of H is obtained by multiplying the number of ml. of N/44 sodium hydroxide by 0.23.

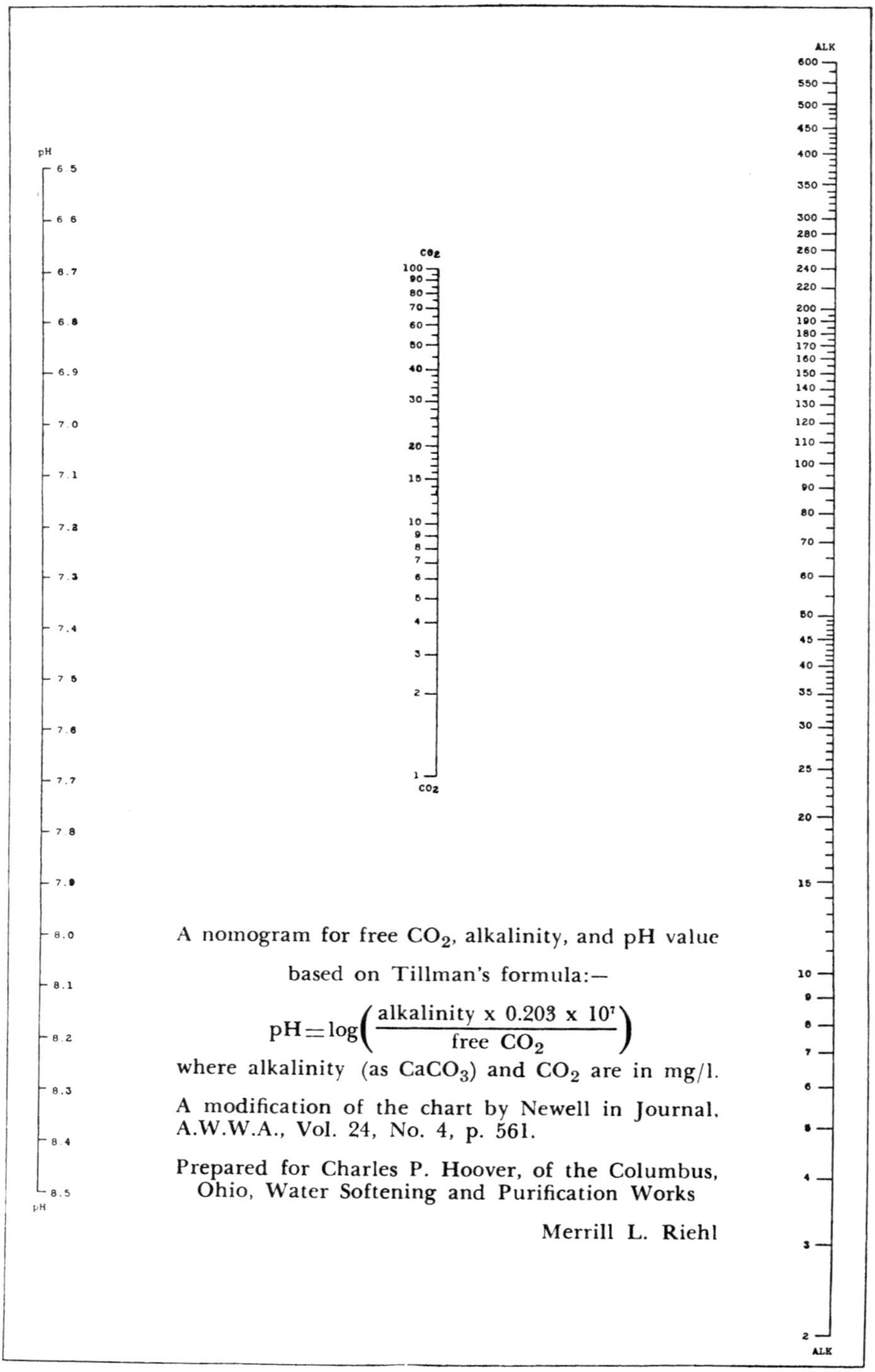


FIGURE 18-1. $CO_2$ NOMOGRAM

*Filterability Index, F.I.*

F.I. is a measure of the ability of treated water to pass through a filter and produce a quality product with minimum head loss and maximum run time. It is a good predictor of pretreatment efficiency and cost analysis, and is much more reliable than settled water turbidity since it does not rely on operator judgement.

To measure the F.I. of a settled water, either from the treatment plant or from the jar tests, obtain a 50–100 ml. sample of settled water just prior to the filters, or from the top portion of the beakers used in the jar tests. If using jar tested water for sampling, take precaution to not disturb the jars in any way. Using the filter funnel which is employed in the bacteriological membrane filter test, place a 0.45 micron membrane filter in the assembly. Wet the filter with a portion of distilled water. With a stopwatch, measure the amount of time it takes to filter 50.0 mls. of room temperature distilled water, and label that time as T1. Next, without disturbing the membrane, pour 50.0 mls. of room temperature settled water through the same filter and measure that time, labeling it as T2. (If it is preferred to measure the settled water at its operating temperature conditions, then the distilled water must be cooled to within 1 or 2 degrees Celsius of the settled water before performing the test).

Calculate F.I. by dividing the settled water filter time, T2, by the distilled water filter time, T1, and record the result. A F.I. of 1.0 is ideal but not achievable. However, values of 1.1 to 1.2 for F.I. are indicative of well treated water which will produce a quality effluent for maximum lengths of filter runs. F.I. values of 1.4 or higher often indicate a need for treatment change. Performed once per shift, F.I. calculations can help to develop a sense of treatment continuity, and will allow trending from which sensible treatment decisions can flow.

*Silica*

*Gravimetric.* Evaporate sample to dryness on a steam bath. Add distilled water and a small quantity of hydrochloric acid. If the nitrates present as $NO_3$ are more than 2 milligrams it is necessary to repeat the evaporation; otherwise filter the residue, wash with hot water, ignite in platinum crucible, cool, and weigh. Add two or three drops of concentrated sulfuric acid, a few ml. of hydrofluoric acid, and heat gently to volatilize the acids. Then ignite and weigh again. The loss of weight is silica ($SiO_2$).

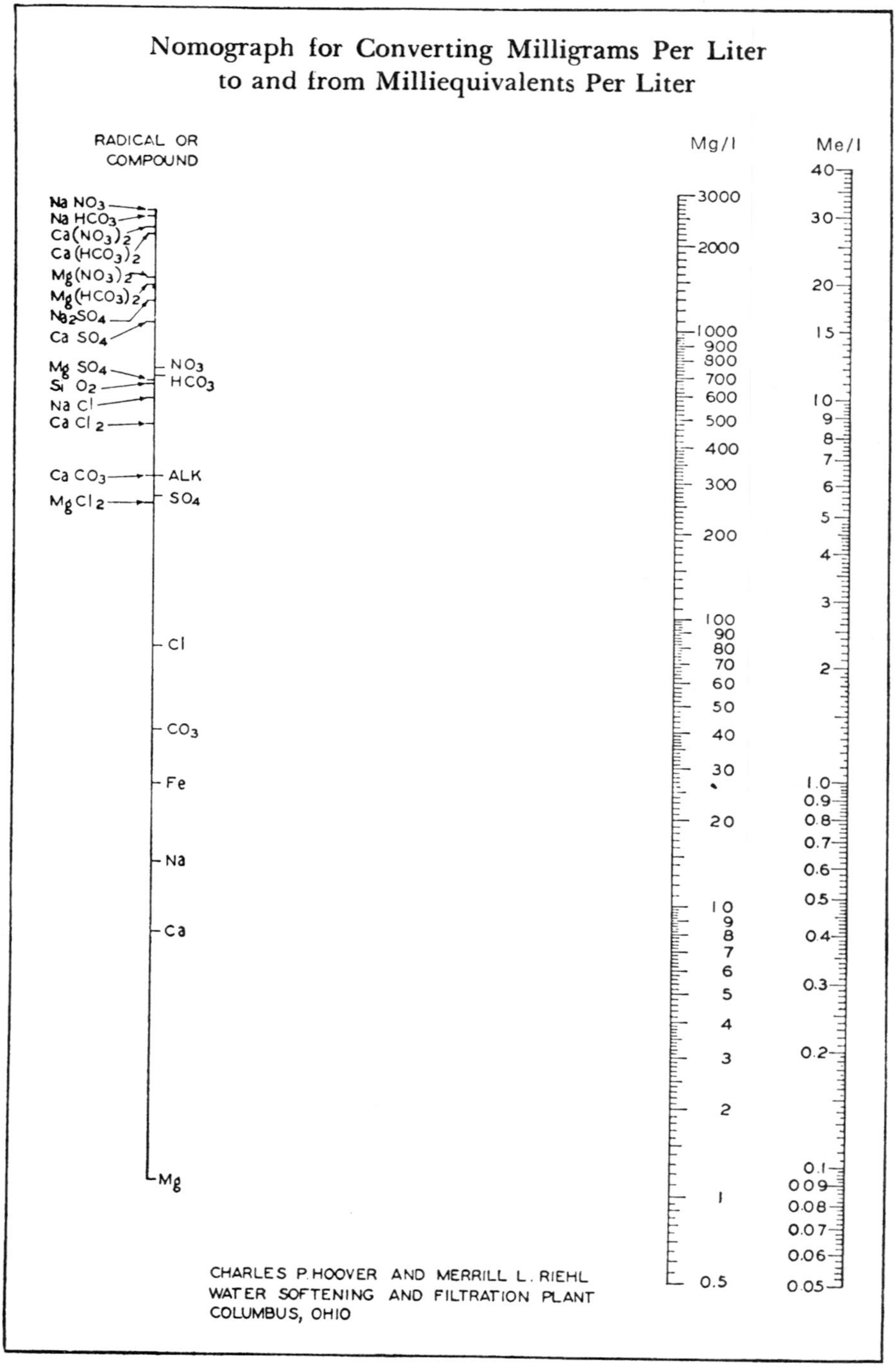


FIGURE 18-2. MILLIEQUIVALENT CONVERSION NOMOGRAPH.

*Alumina*

Reagents.

1. Acetic acid, 30 percent.
2. Ammonium carbonate, saturated.
3. Hematoxylin solution, dissolve 0.1 grams of C.P. hematoxylin in 100 ml. of boiling distilled water or 100 ml. pure ethyl alcohol. This solution is stable for two or three weeks.
4. Standard aluminum sulfate solution, $Al_2(SO_4)_3 \cdot 18H_2O$, 0.3269 grams per liter. (Commercial aluminum sulfate contains about $14H_20$.) It may be necessary to make this solution N/100 when the $Al_2O_3$ content of the sample is low.

*Procedure.* Place 50 ml. of the water to be tested in a 50 ml. tall form Nessler tube. Adjust pH to 4.5 to 5.0 (end point for methyl orange indicator). Add 1 ml. of saturated solution of ammonium carbonate and 1 ml. of hematoxylin solution. Mix by inverting the tube. Allow to stand 20 minutes in order to develop the maximum color. Make up standards in the same manner. Add 1 ml. of 30 percent acetic acid to each tube and compare. Calculation: 1 ml. $Al_2(SO_4)_3 \cdot 18H_2O$ standard in 50 ml. water = 1 mg/l $Al_2O_3$.

*Sodium Polyphosphate*

Reagents

1. 4N sulfuric acid—120 ml. of concentrated acid per liter.
2. Standard potassium phosphate (monobasic) solution—0.1335 grams of $KH_2PO_4$, reagent grade, per liter. 1 ml. per 100 ml. = 1 mg/l.
3. 5 percent ammonium molybdate—50 grams, reagent grade, per liter. Filter concentrated solution if cloudy. Add 50 ml. concentrated nitric acid (C.P.).
4. 0.05 percent stannous chloride—make fresh daily. Titrate 0.05 grams of C.P. stannous chloride with 0.6 ml. of concentrated hydrochloric acid in a mortar, filter through fine filter, and make up to 100 ml. with distilled water.

*Procedure.*

1. Treat 100 ml. of the sample or samples with 5 ml. of 4N sulfuric acid, including a blank on the untreated or polyphosphate-free sample.
2. Boil down to approximately 10 ml. and cool.
3. Make up standards, using 0.2, 0.4, 0.6, etc., ml. of the $KH_2PO_4$ standard solution in the range desired.

4. Treat the standards and an unreverted (not boiled) blank with 5 ml. of 4N sulfuric acid and make standards up to 50 ml. with untreated water in order to have standards and samples of similar composition.
5. Treat standards, unreverted blank, and reverted samples with 2 ml. of ammonium molybdate solution. Run an unreverted blank on each water of different composition or characteristics.
6. Treat all with 5 ml. of 0.05 percent stannous chloride solution.
7. Make up standards to 100 ml. with untreated water and samples with distilled water, and observe at 5 minutes.
8. Polyphosphate = reverted determination minus unreverted determination.
9. pH of Columbus, Ohio, tap water (lime-soda ash softened) with this treatment = 8.
10. If concentration of polyphosphate is greater than 1 mg/1 use a smaller sample; i.e., if the concentration is 2 mg/1 use 25 ml. of sample diluted to 100 ml. with phosphate-free, untreated water, etc. The reason for this is that at 1 mg/1 the blue color becomes too dark to read.

*Reporting Results of Water Analysis*

Alkalinity and hardness results are reported in terms of $CaCO_3$. Silicon (Si) is expressed as $SiO_2$. The bases and acids or positive and negative radicals are reported in ionic forms.

The results so reported show actual determinations and permit other chemists to make hypothetical combinations according to their own choice of method. It is also possible from results reported in this way to judge the probable accuracy of the work. The accuracy of an analysis may be estimated by comparing the sum of the milliequivalents per liter (me/l) of the positive radicals (cations) with the sum of the milliequivalents per liter of the negative radicals (anions). In a perfect analysis they would be exactly the same. The percentage of error may be figured readily by the Stabler formula:

e = rp–rn/rp+rn x 100

e = percentage of error.

rp = sum of me/l of positive radicals.

rn = sum of me/l of negative radicals.

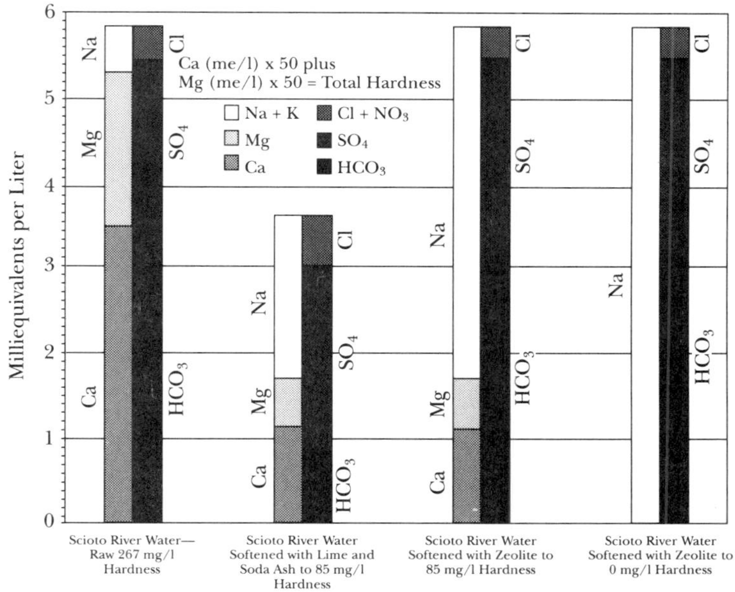


The above diagram shows the Scioto River water at Columbus, Ohio (1) before treatment, (2) after coagulation with aluminum sulphate and softening with lime and soda ash, and (3–4) softening by ion exchange.

FIGURE 18-3. DIAGRAM OF SCIOTO RIVER TREATMENT

The milliequivalents per liter are determined as follows: me/l = mg/l of element or compound × Valence/Atomic Weight (found by analysis)

Conversely, if results are expressed as milliequivalents per liter, milligrams per liter are determined as follows:

mg/l = me/l × Atomic Weight/Valence

The following table shows the coefficients for converting milligrams per liter (mg/l) into milliequivalents per liter (me/l):

| *Positive radicals* | | | |
|---|---|---|---|
| mg/l Iron | (Fe) | × | 0.0358=me/l |
| Mg/l Aluminum | (Al) | × | 0.1112=me/l |
| mg/l Calcium | (Ca) | × | 0.0499=me/l |
| mg/l Magnesium | (Mg) | × | 0.0822=me/l |
| mg/l Sodium | (Na) | × | 0.0435=me/l |
| mg/l Potassium | (K) | × | 0.0256=me/l |
| mg/l Manganese | (Mn) | × | 0.0364=me/l |
| mg/l Hydrogen | (H)* | × | 0.9921=me/l |

* Present in acid water only.

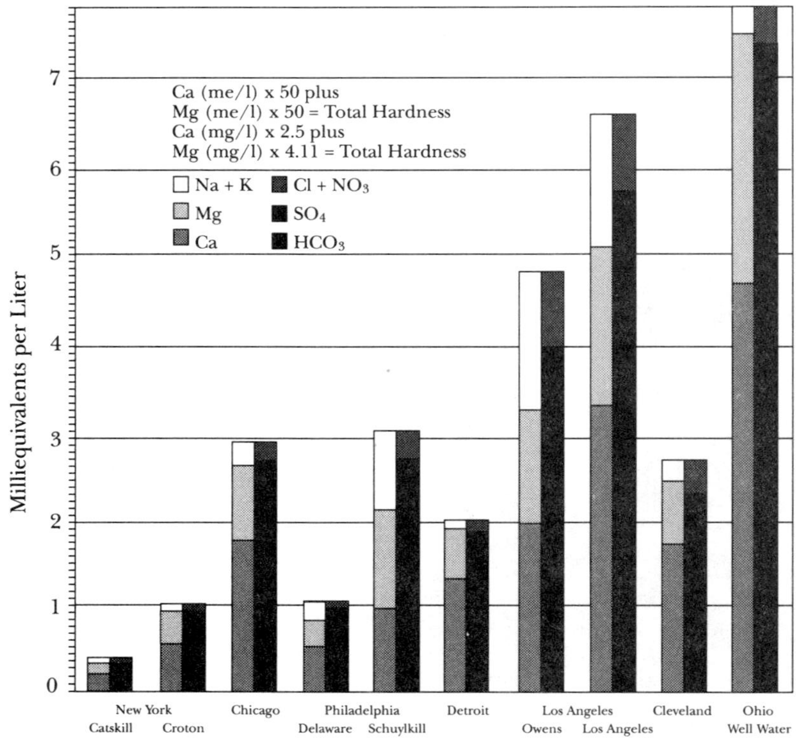


Composition of water at the six largest cities in the United States and the largest public well supply in Ohio.

FIGURE 18-4. DIAGRAM OF COMPOSITION OF WATER AT VARIOUS LOCATIONS

*Negative Radicals*

| | | | |
|---|---|---|---|
| mg/l Carbonate | ($CO_3$) | × | 0.0333=me/l |
| mg/l Bicarbonate | ($HCO_3$)* | × | 0.0164=me/l |
| mg/l Sulfate | ($SO_4$) | × | 0.0208=me/l |
| mg/l Chloride | (Cl) | × | 0.0282=me/l |
| mg/l Fluoride | (F) | × | 0.0526=me/l |
| mg/l Nitrate | ($NO_3$) | × | 0.0161=me/l |

* Bicarbonate Alkalinity expressed as $CaCO_3$ x 1.22; or 1.22; or 1.22 × (T-2P).

The following table shows the coefficients for converting me/l into mg/l.

*Positive Radicals*

| | | | |
|---|---|---|---|
| me/l Iron | (Fe) | × | 27.92=mg/l |
| me/l Aluminum | (Al) | × | 8.99=mg/l |
| me/l Calcium | (Ca) | × | 20.04=mg/l |
| me/l Magnesium | (Mg) | × | 12.16=mg/l |
| me/l Sodium | (Na) | × | 23.00=mg/l |
| me/l Potassium | (K) | × | 39.10=mg/l |
| me/l Manganese | (Mn) | × | 27.46=mg/l |
| me/l Hydrogen | (H) | × | 1.01=mg/l |

*Negative Radicals*

| | | | |
|---|---|---|---|
| me/l Carbonate | ($CO_3$) | × | 30.00=mg/l |
| me/l Bicarbonate | ($HCO_3$) | × | 61.01=mg/l |
| me/l Sulfate | ($SO_4$) | × | 48.03=mg/l |
| me/l Chloride | (Cl) | × | 35.46=mg/l |
| me/l Fluoride | (F) | × | 19.00=mg/l |
| me/l Nitrate | ($NO_3$) | × | 62.01=mg/l |

Reporting results in terms of positive and negative radicals does not always satisfy the layman for it gives him a hazy impression of the composition of the water; therefore the results may be reported according to a graphic scheme or a probable hypothetical combination.

For convenience in converting mg/l to and from me/l, see nomograph earlier in this chapter.

## *Graphic Scheme*

The milliequivalents per liter may be considered as an arbitrary figure which for the purposes of plotting may be called inches or centimeters. That is each me/l = 1 inch or 1 centimeter. If a strip of paper is cut for each positive and each negative ion, the length corresponding to the equivalent expressed in inches or centimeters, and these strips laid in two rows, the cations on the top row and the anions on the under row, and arranged according to a recognized system of pairing the positive and negative radicals used for calculating probable hypothetical combinations, a graphical representative will be obtained, or graphical diagram may be drawn. It is customary in some laboratories to show calcium, magnesium, sodium, potassium, bicarbonate, sulfate, chloride, and nitrate diagrammatically.

The following is a typical example:

*Analysis of Water*

| | | |
|---|---|---|
| Ca 100 mg/l × | 0.0499 = approximately 5.0 me/l | |
| Mg 30 mg/l × | 0.0822 = approximately 2.5 me/l | |
| Na 5 mg/l × | 0.0435 = approximately 0.2 me/l | |
| $HCO_3$ 244 mg/l × | 0.0164 = approximately 4.0 me/l | |
| $SO_4$ 159 mg/l × | 0.0208 = approximately 3.3 me/l | |
| Cl 14 mg/l × | 0.0282 = approximately 0.4 me/l | |

In drawing the picture, arrange the equivalents of the analysis as follows:

Top row—Calcium, Magnesium, and Sodium. Bottom row—Bicarbonate, Sulfate, and Chloride. The results of the positive and negative ions may not exactly equal each other, in which case the errors of closure may be distributed between the calcium and carbonate ions.

For a single analysis it may be convenient to show the diagram arranged horizontally. If, however, a series of analyses are to be shown it is preferable to arrange them vertically according to the method used for a number of years by the United States Geological Survey, and explained in its Water Supply Paper 596-H.

Results shown diagrammatically. Each me/l = 1 centimeter.

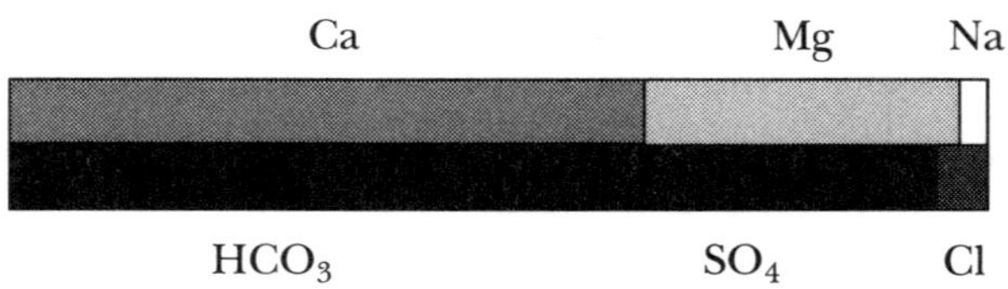


## *Calculating Hypothetical Combinations*

The method of Hale for combining ions is recommended. The scheme is based on the theory that the total alkalinity first should be calculated to calcium. If the carbonates exceed the required amount of calcium, the rest of the alkalinity may be calculated to magnesium carbonate. Sulfates, chlorides, and similarly radicals are combined with magnesium until the latter is exhausted. If the calcium exceeds the amount necessary to account for the alkalinity, there is no magnesium carbonate, and the excess of calcium may be calculated to sulfates first, then to chlorides and nitrates until exhausted, and the magnesium is combined with the remaining acid radicals until exhausted. Any remaining chlorides, nitrates, or sulfates may be calculated to sodium and potassium salts.

An examination of the horizontal graph shows that the water contains calcium bicarbonate*, calcium sulfate, magnesium sulfate, magnesium chloride, and sodium chloride. If it is desired to find out approximately how much of these compounds is present it can be done readily by measuring the various compounds present with a ruler, remembering that each 1 centimeter = 1 me/l. The me/l can then be converted to mg/l by multiplying by the factors given or by use of the nomograph.

For example:

4.0 me/l or 80 mg/l Ca are present as $HCO_3$
1.0 me/l or 20 mg/l Ca are present as $SO_4$
2.3 me/l or 28 mg/l Mg are present as $SO_4$
0.2 me/l or 2 mg/l Mg are present as Cl
0.2 me/l or 5 mg/l Na are present as Cl

By multiplying by the factors given in the following table, the mg/L of the compounds present may be obtained:

| *Found* | *Sought* | *Factor* |
|---|---|---|
| Ca | $CaCO_3$ | 2.50 |
| Ca | $CaSO_4$ | 3.40 |
| Ca | $CaCL_2$ | 2.77 |
| Mg | $MgCO_3$ | 3.47 |
| Mg | $MgSO_4$ | 4.95 |
| Mg | $MgCl_2$ | 3.92 |
| Na | $Na_2CO_3$ | 2.30 |
| Na | $Na_2SO_4$ | 3.09 |
| Na | NaCl | 2.54 |

* In reporting the bicarbonate, some chemists prefer to call the compound carbonate; i.e., the form in which it is determined on evaporation, although in solution in natural waters it is logically considered to be as bicarbonate. As far as the me/l is concerned, the value is the same for either form, the only distinction being the choice of factors to use in converting to mg/l.

## Calcium Carbonate Chemical Balance or Stability Test

1. Fill two 500 ml. glass stopper bottles completely full of water to be tested. Let bottles be subjected to same temperature and time conditions.
2. Add about 1/2 gm. of reagent grade precipitated $CaCO_3$ powder to one. Place stoppers in both bottles.
3. Mix by inverting every five minutes for 1/2 hour or longer, or else use a magnetic type of stirrer.

4. Filter both through clean filter paper into clean beakers.
5. Determine Methyl orange alkalinity on each. Results on untreated sample designated as (a) and on calcium carbonate treated sample as (b).
6. If (a) is greater than (b) the water is supersaturated with carbonate and may be scale forming.
7. If (a) is less than (b) the water is undersaturated as to carbonate and may be corrosive.
8. If (a) is equal to (b) the water is in equilibrium as to carbonate constituents.

W. F. Langelier, Associate Professor of Engineering, University of California, developed a formula* whereby the pH saturation of the water can be computed from the results of a water analysis. Knowing the total dissolved solids, the calcium, alkalinity, and temperature, the pH saturation can be calculated. The difference between the calculated pH saturation and the actual pH is expressed as the saturation index. See graph and nomogram on following page.

*Graph/Nomogram*

If it is desired to use pH values instead of alkalinity determinations, and to report the results in terms of saturation indices as suggested by Langelier, proceed as follows:

1. Determine the pH value on a portion of the sample.
2. To the remainder of the sample add C.P. precipitated calcium carbonate in excess, stir a few minutes, allow to settle, and filter.
3. (a) If the pH value in (2) is greater than in (1) it shows that the water is undersaturated as to carbonate constituents and may be corrosive. The difference between the pH values is the saturation index. For example, if the sample before the addition of $CaCO_3$ has a pH value of 7.4, and after the addition of $CaCO_3$, 8.3, the saturation index would be –0.9.
   (b) If the pH values in (1) and (2) are equal, the water is in equilibrium as to carbonate constituents.
   (c) If the pH value in (1) is greater than in (2), the water is supersaturated with carbonate compounds and may be deposit films in piping, etc. If the pH on (1) is 9.3 and on (2) is 9.0, then the saturation index is +0.3.

* Journal, A.W.W.A., Vol. 28, October, 1936.

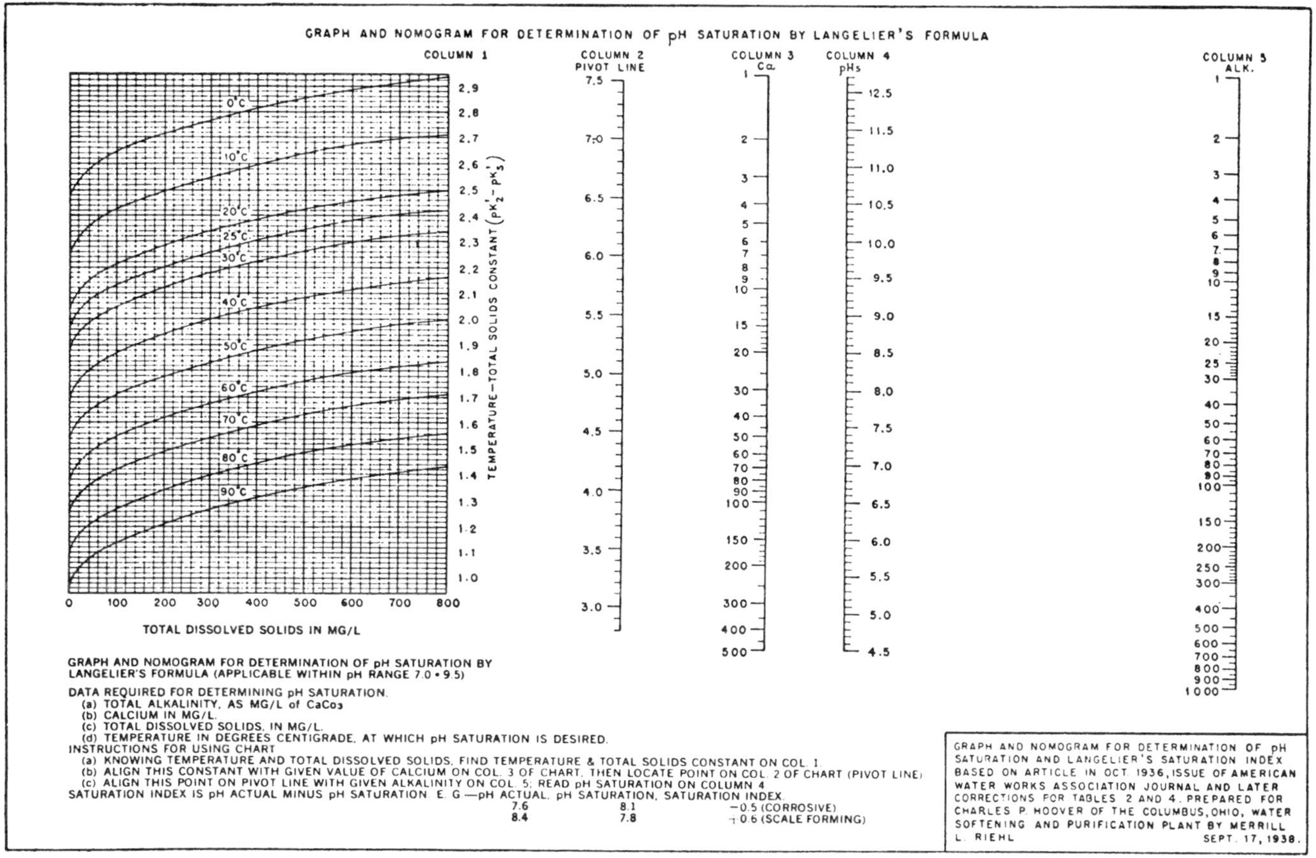


FIGURE 18-5. LANGELIER'S FORMULA

*Quantitative Determination of Odor*

The "threshold odor value" or T.O.V. is measured as follows:

Apparatus and reagents

1. Granular carbon "filter" for preparation of odor-free water
2. Hot plate or water bath
3. Glass stoppered Erlenmeyer flasks
4. Graduated cylinders 25, 100, and 250 ml.
5. Mohr type pipettes 5 and 10 ml., graduated to 0.1 ml.
6. Thermometer 0–100° C.
7. Sample bottles

*Procedure:*

1. Cleansing of glassware; this must be thoroughly clean, and rinsed with hot odor-free water. Avoid use of cleansing agents with perfumes, or odor-masking compounds.
2. Preparation of odor-free water. Slowly pass a supply of water, preferable having little odor originally, thru the column of activated carbon. Waste first effluent, until water is assured to be odor free.
3. Temperature of measurements. Odor measurements have generally been made at about 20°C. and 60°C., the higher temperature being used to release the more volatile odor bearing material. However, at 60° the olfactory nerves may be dulled and better results may be reached with measurements at 40°C.
4. Dilute a portion of odor-free water; place in glass stoppered flask, heat to 40°, shake rather briskly, remove stopper, and note whether there is an odor in this dilution as compared to odor-free water "blank" that is treated in the same way. If an odor was noted, repeat with another dilution using less sample; but if no odor difference was detectable, more sample is taken for the second step.
5. Calculations: The smallest amount of sample taken in which odor could be observed, is used to compute the results by this formula:

   T.O.V. = 200/ml. Sample

*Outline of Laboratory Odor Experiments*

*Break-Point Chlorination Treatment.* Make a solution of chlorine by adding 10 ml. of Zonite to 100 ml. distilled water. (Solution so made contains 1000 mg/l, or 1 gram per liter, of available chlorine.) 1 ml. to

1000 ml. water equals 1 mg/l. Or, a solution of chlorine water (1 gram per liter) may be made by passing chlorine gas from a chlorine cylinder into distilled water. Check strength by titrating with N/10 sodium thiosulfate. Treat a series of samples with chlorine in increments of 1 mg/l beginning with 1 milligram. Note the sample in which an instantaneous yellow color is developed. The amount of chlorine required to produce this yellow color is the flash point which usually occurs at or closely past the breakpoint.

In order to determine whether or not the chlorine treatment has eliminated the odor from the water, dechlorinate. Absence of chlorine should be checked by orthotolidine before making test. After complete dechlorination, filter and determine threshold odor value and taste.

Sulfur dioxide gas solution, preferably N/100, may be prepared as follows: Generate $SO_2$ gas by gently heating a slightly acid (HCl) solution of sodium sulfite or sodium bisulfite, and absorbing the gas in distilled water. Standardize the $SO_2$ solution by titrating with N/100 iodine solution. In dechlorinating with $SO_2$ gas it is convenient to use N/100 solution because each ml. when added to 355 ml. of sample will neutralize exactly 1 mg/l of excess chlorine.

*Carbon Treatment.* Coagulate a large sample with alum, if water is turbid, and let settle. Divide this sample into smaller portions and add carbon at the rate of 10, 20, 30, 50, 100, 150, 200, 250, and 300 pounds per million gallons, stir 30 minutes, filter samples through paper filters, and determine odor.

Repeat above experiment using both alum and lime.

Repeat above experiment, first aerating water.

*Permanganate Treatment.* Treat samples with varying amounts of permanganate, then follow this treatment with lime treatment, carrying a causticity of 20 to 30 mg/l. Determine the maximum quantity of permanganate that can be added without leaving residual manganese in the water after lime treatment; i.e., with 20 or 30 mg/l excess lime. Make odor tests on the permanganate treated samples after lime treatment and filtration through paper filters.

### *Procedure for Determination of Algae and Microorganisms*

The operation of a plant treating surface supply can combat tastes and odors due to algae and similar organisms more effectively if regular examination of the supply is made to determine the type and number of organisms present.

Such tests are not difficult, the following apparatus being needed:

1. Microscope, with lens system giving 75 power (diameters) magnification
2. Whipple disk for eyepiece
3. Counting cell
4. Segwich Rafter Funnel, with stand
5. Aspirator (suction pump)
6. Reference charts showing pictures of various organisms

*Procedure:*

1. Preparation of Segwich Rafter Funnel:The silk bolting cloth is place over the one hole stopper which is then inserted in the bottom of the funnel. About 1 ml. of Berkshire sand is poured into the funnel, and the suction hose placed on the glass tube leading from the rubber stopper.
2. Gentle suction is applied and the water sample (one liter or less) is pulled thru the sand, increasing suction as needed.
3. Carefully remove the stopper, allowing sand to drop into a small beaker. 2 or 3 ml. of distilled water is used to rinse the remaining sand grains from the small neck of the funnel. The bolting cloth and stopper are then held over the beaker and 2 or 3 ml. of water used to rinse these. The organisms are suspended in the small volume of water by swirling, then after the sand settles for a few seconds, the water is decanted to a 10 ml. graduate. The sand is given a second rinsing with 2 or 3 ml. of distilled water, which is also poured into the graduate, and the total volume brought to 10 ml. If a 1 liter sample were used, the organisms have been concentrated by 1000/10 or 100 times.
4. Mix cylinder by inverting, then pipette a portion into the counting cell. In this step the cover is left on, but turned slightly with two corners of the cell open, the sample being inserted in one corner, the air escaping from the other. Slide cover in place, excluding all air bubbles, and place on the microscope, which has been set for exactly 75 diameters magnification.
5. The total area within the area of the Whipple disk is 1 mm. square, and since the cell is 1 mm. deep, all the organisms in 1 cubic millimeter of the concentrate (100 cubic mm. of the original sample) can be counted. Record the total number of each type of organisms for 10 fields selected at random; this sum multiplied by 1000 gives the count per liter.

Chapter XIX

# Miscellaneous Analyses, Tests, and Standards

## *Analysis of Lime*

The most widely used test for determining available lime is the Rapid Sugar test, which is specified by two important national organizations, the American Water Works Ass'n. and the American Society for Testing and Materials:

AWWA Standard for Quicklime and Hydrated Lime B 202-93
ASTM Chemical Analysis of Lime C-25.

**Rapid Sugar Test.** Dissolve 0.5 gram of -#100 mesh sample of lime in 170 ml. of a 10 percent sugar solution, shake, and let stand for 30 to 60 minutes. Then titrate with 0.1782N HCl solution, using phenolphthalein indicator. 1 ml. of this standard acid solution is equivalent to 1% available CaO.

A modification of the Rapid Sugar test is used in which N/10 sulfuric acid is used in place of HCl. In this test the burette reading, after titration, multiplied by 2.244 = percent CaO.

*Complete Chemical Analysis.* The methods of the American Society for Testing and Materials (C 25) are recommended.

*Slaking Rate Test.* The slaking rate of quicklime is a measure of its reactivity, which is determined by a test method included in the appropriate AWWA standard: B-202-93 (adapted from ASTM C110). The quicklime is slaked in a dewar flask fitted with a mechanical stirrer and thermometer, and the temperature rise is recorded at 30- and 60-second intervals.

The results are reported as temperature rise in 3 minutes, total temperature rise, and total active slaking time.

## *Specifications for Lime*

Either quicklime or hydrated lime of high calcium content can be used for water treatment. Specifications for quicklime usually call for a minimum of 90 percent available CaO and for hydrated lime, a minimum of 68 percent available CaO. Complete specifications for lime for water treatment can be found in the latest AWWA Standard B202-93.

### *Analysis of Soda Ash*

Dissolve 2 grams sample in 200 ml. cold, recently boiled distilled water. Titrate 50 ml. with N/2 sulfuric acid, using methyl orange as indicator. Burette reading multiplied by 5.3 = percent $Na_2CO_3$.

### *Specification for Soda Ash*

The soda ash shall be that known as 58 percent soda ash, and shall contain not less than 98 percent sodium carbonate. The material shall be in in a dry powdered form, shall contain no large lumps or large crystals.

### *Specification for Aluminum Sulfate*

The material shall be basic and shall contain water soluble aluminum sulfate equivalent to not less than 17 percent alumina ($Al_2O_3$). Percentage of insoluble material shall be stated by bidder.

### *Analysis of Aluminum Sulfate*

Insoluble Matter. Dissolve 5 grams of sample and digest for 5 minutes with 100 ml. water. Filter into 500 ml. volumetric flask, wash the insoluble residue, dry, blast for 15 minutes, and weigh. Make up filtrate to exactly 500 ml.

*Total Iron.* Dissolve 10-gram sample in 100 ml. distilled water, then filter. Reduce with hydrogen by means of zinc and sulfuric acid. Titrate with a dilute standard solution of $KMnO_4$. Report as $Fe_2O_3$.

*Total Alumina.* Take 50 ml. of filtrate (0.5 gram sample) and dilute to about 150 ml. Add 2 ml. HCl and a few drops of $HNO_3$, and heat to boiling. Make slightly alkaline with $NH_4OH$ and boil five minutes. Filter, wash four times, ignite by blasting 20 minutes, weigh as $R_2O_3$, and calculate to percentage. $Al_2O_3=R_2O_3$ minus total Fe as $Fe_2O_3$.

*Free Alumina or Free Sulfuric Acid.* Take 50 ml. of filtrate from insoluble residue, dilute to 300 ml., and heat to boiling. Add 1 ml. phenolphthalein and titrate with a normal solution of NaOH to a faint pink color. Again boil 2 minutes and titrate, and repeat until the pink color remains upon boiling.

Calculate the soda used in terms of $H_2SO_4$. (1 ml.1 NaOH solution is equivalent to 0.04904 gram $H_2SO_4$.)

$H_2SO_4$ equivalent to $Al_2O_3$ found = $Al_2O_3 \times 2.8862$.
$H_2SO_4$ equivalent to $Fe_2O_3$ found = $Fe_2O_3 \times 1.8426$.

If the sulfuric acid found by titration is greater than found equivalent to the bases, the excess is due to free sulfuric acid. If the sulfuric acid by titration is less than the equivalent, the difference divided by 2.8792 equals free or basic $Al_2O_3$.

*Water.* The percentage of water is found by difference. Add together the percentages of $SO_3$ ($H_2SO_4$ x 0.8163=$SO_3$), insoluble matter, iron and aluminum oxides, and subtract total from 100.

## *Analysis of Filter Sand for Effective Size and Uniformity Coefficient*

The sand sample taken for analysis should not be a random sample, but should be representative of the entire quantity, and should be thoroughly dry before being tested.

A series of standard sieves, preferable from a calibrated set, is selected to cover several sizes larger and several sizes smaller than the approximate "effective size". A carefully weighed sample of from 100 to 200 grams is poured into the top (largest opening) sieve, and shaken thoroughly by hand or machine for at least 5 minutes. The sand remaining on each sieve is carefully removed, including the grains caught in the mesh, and weighed to the closest 0.1 gram. The grams passing, and % of total weight passing each sieve is computed, as illustrated by the following table:

| | *Sieve No. and Nominal size* | *Sieve opening im Millimeters* | *Grams on the sieve* | *Grams Passing* | *% Passing* |
|---|---|---|---|---|---|
| E | 18 | 1.08 | 0 | 119.2 | 100 |
| F | 20 | 0.83 | 1.1 | 118.1 | 99.1 |
| G | 25 | 0.73 | 13.7 | 104.4 | 87.6 |
| H | 30 | 0.64 | 42.2 | 62.2 | 52.2 |
| I | 35 | 0.50 | 43.3 | 18.9 | 15.9 |
| J | 40 | 0.418 | 15.2 | 3.7 | 3.1 |
| K | 45 | 0.358 | 2.9 | 0.8 | 0.7 |
| Pan | — | 0 | 0.8 | 0 | 0 |
| | Total | | 119.2 | | |

The data on % passing are transferred to a chart (log-log graph paper, usually specially prepared for the sieves) and a smooth curve drawn through the plotted points. From the curve obtained, pick the points where the line crosses the 10% passing and the 60% passing lines (Points A and B). Read the millimeters opening size which correspond to these points.

The value for "A" is the effective size, while the value for "B" is divided by the value for "A" to give the uniformity coefficient. In the curve obtained, in the chart shown, "A" was found to be 0.46 mm., and "B" to be 0.66 mm. Hence the effective size is 0.46 mm., and the uniformity coefficient is 6.00/0.46 or 1.44.

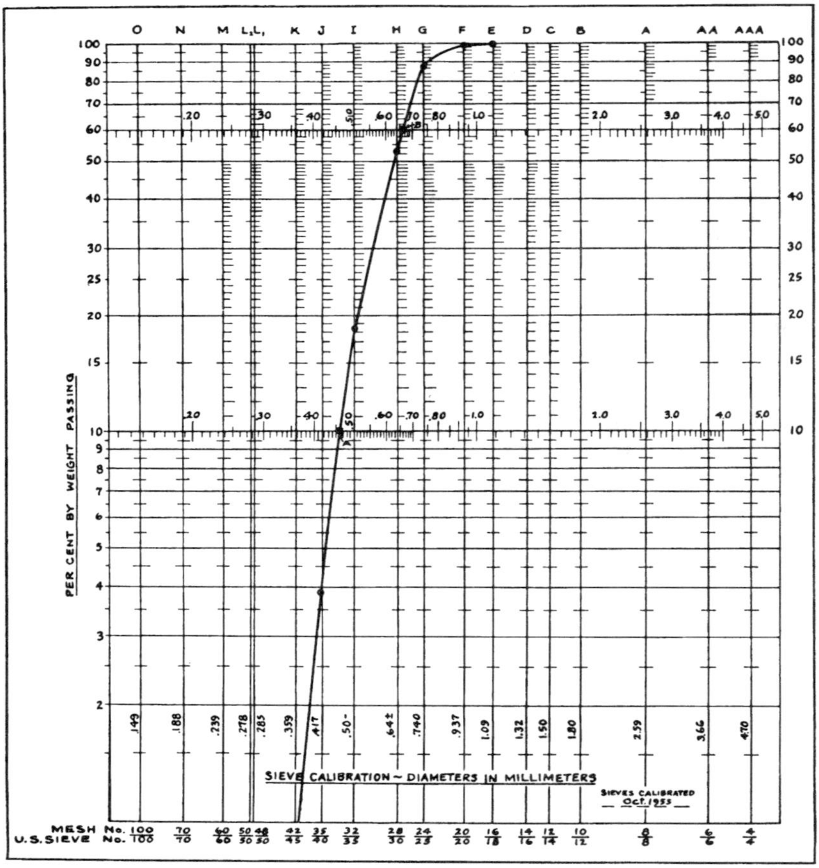


FIGURE 19-1. FILTER SAND CHART

*Preparation of Reagents for Laboratory Experiments*

For dose of grains per gallon added to gallon samples, use 64.7 grams of chemical to liter. 1 ml. to 1 gallon = 1 grain per gallon.

For dose of milligrams per liter added to gallon samples, use 3.8 grams of chemical in 1 liter. 1 ml. to 1 gallon = 1 mg/l.

For dose of pounds per million gallons added to gallon samples, use 0.454 grams of chemical to 1 liter. Then add 1 ml. to 1 gallon=1 pound per million gallons. Smaller doses than 10 pounds per million gallons are seldom used, therefore it is better to make stock solution 10 times this strength so that 1 ml. = 10 pounds per million gallons.

For dose of grains per gallon added to 1 liter samples, use 17.1 grams of chemical to 1 liter. Then add 1 ml. to 1 liter = 1 grain per gallon.

For dose of milligrams per liter added to 1 liter samples, use 1 gram of chemical to 1 liter. Then add 1 ml. to 1 liter = 1 mg/l.

For dose of pounds per million gallons added to 1 liter samples use 0.12 grams of chemical per liter. Then add 1 ml. to 1 liter = 1 pound per million gallons. Smaller doses than 10 pounds per million gallons are seldom used, therefore it is better to make this stock solution 10 times this strength so that 1 ml. = 10 pounds per million gallons.

### *The Safe Drinking Water Act (SDWA)*

The SDWA was passed in 1974, and amended in 1986, and it requires USEPA to publish and regulate drinking water standards that apply to public water systems. The Act specifies those contaminants that may have an adverse health effect on persons, and requires all water systems serving 25 or more people to achieve compliance with maximum contaminant levels (MCLs), or provide treatment techniques for those contaminants. The term maximum contaminant level means the maximum allowable level of a contaminant in water that is delivered to any user of a public water system. MCLs are enforceable, but EPA is allowed to make variances for those systems which have demonstrated best available technology (BAT).

### *Maximum Contaminant Level (MCL) Standards for Public Water Supplies*

**Effective Date January 1, 1993**

**Chemical Standards**

| *Inorganic* | *MCL (mg/L)* |
|---|---|
| Antimony | 0.006 |
| Arsenic | 0.05 |
| Asbestos | 7 million fibers/liter (longer than 1 um) |
| Barium | 2. |
| Beryllium | 0.004 |
| Cadmium | 0.005 |

**Chemical Standards** *(continued)*

| *Inorganic* | *MCL (mg/L)* |
|---|---|
| Cyanide | 0.2 |
| Chromium | 0.01 |
| Fluoride | 4.0 |
| Lead | 0.05 |
| Mercury | 0.002 |
| Nickel | 0.1 |
| Nitrate (as N) | 10. |
| Nitrate-Nitrite (as N) | 10. |
| Nitrite (as N) | 1. |
| Selenium | 0.05 |
| Thallium | 0.02 |

**Pesticides and Other Organic**

| *Chemicals (SOCs)* | *MCL (mg/L)* |
|---|---|
| Endrin | 0.002 |
| Lindane | 0.0002 |
| Methoxychlor | 0.04 |
| Toxaphene | 0.003 |
| 2,4-D | 0.07 |
| 2,4,5-TP Silvex | 0.05 |
| Alachlor | 0.002 |
| Atrazine | 0.003 |
| Carbofuran | 0.04 |
| Chlordane | 0.002 |
| Dibromochloropropane (DBCP) | 0.0002 |
| Ethylene dibromide (EDB) | 0.00005 |
| Heptachlor | 0.0004 |
| Heptachlor epoxide | 0.0002 |
| PCB's | 0.0005 |
| Pentachlorophenol | 0.001 |
| Dalapon | 0.2 |
| Dinoseb | 0.007 |
| Diquat | 0.02 |
| Endothall | 0.1 |
| Glyphosate | 0.7 |
| Oxamyl (Vydate) | 0.2 |

**Pesticides and other organic** (*continued*)

| *Chemicals (SOCs)* | *MCL (mg/L)* |
|---|---|
| Picloram | 0.5 |
| Simazine | 0.004 |
| Benzo-a-pyrenes | 0.0002 |
| Di(2-ethylhexyl)adipate | 0.4 |
| Di(2-ethylhexyl)phthalate | 0.006 |
| Hexachlorobenzene | 0.001 |
| Hexachlorocyclopentadiene | 0.05 |
| Dioxin | 0.00000003 |
| Aldicarb | 0.003 (proposed) |
| Aldicarb sulfoxide | 0.004 (proposed) |
| Aldicarb sulfone | 0.002 (proposed) |
| Metolachlor | not yet established |
| Acrylamide | treatment technique |
| Epichlorohydrin | treatment technique |

| *Trihalomethanes (THMs)* | *MCL (mg/L)* |
|---|---|
| Total trihalomethanes: | 0.10 |

| *Volatile Organic Chemicals (VOCs)* | *MCL (mg/L)* |
|---|---|
| Benzene | 0.005 |
| Vinyl chloride | 0.002 |
| Carbon tetrachloride | 0.005 |
| 1,2-Dichloroethane | 0.005 |
| Trichloroethylene | 0.005 |
| para-Dichlorobenzene | 0.075 |
| 1,1,-Dichloroethane | 0.007 |
| 1,1,1-Trichloroethan | 0.20 |
| o-Dichlorobenzene | 0.6 |
| cis-1,2-Dichloroethylene | 0.07 |
| trans-1,2-Dichloroethylene | 0.1 |
| 1,2-Dichloropropane | 0.005 |
| Ethylbenzene | 0.7 |
| Monochlorobenzene | 0.1 |
| Styrene | 0.1 |
| Tetrachloroethylene | 0.005 |

**Pesticides and other organic** (*continued*)

| *Volatile Organic Chemicals (VOCs)* | *MCL (mg/L)* |
|---|---|
| Toluene | 1. |
| Xylenes (total) | 10. |
| Dichloromethane | 0.005 |
| 1,2,4-Trichlorobenzene | 0.07 |
| 1,1,2-Trichloroethane | 0.005 |

| *Radiological* | *MCL (pico curie/L)* |
|---|---|
| Combined Radium 226 and Radium 228 | 5 |
| Gross Alpha particle activity (including Radium 226 but excluding Radon and Uranium) | 15 |
| Beta particle | 50 |
| Tritium | 20,000 |
| Strontium 90 | 8 |

| *Microbiological Standards Presence/Absence Total Coliform* | |
|---|---|
| Small system | 1 positive sample/month |
| Large system | 5% positives/month |

The Surface Water Treatment Rule (SWTR) of the SDWA provides that turbidity shall be limited to 0.5 NTU in 95% of the samples taken at the plant tap in a given month, and that 99.9% removal of Giardia cysts and 99.99% removal of viruses must be achieved.

## Useful Equivalents

1 gallon of water weighs 8.34 pounds
1 cubic foot equals 7.48 gallons
1 gallon equals 3,785 milliliters
1 grain per gallon equals 17.1 mg/L
1 grain per gallon equals 142.9 pounds per million gallons
1 mg/L equals 8.34 pounds per M.G.
1 M.G.D. equals 694 gpm
1 M.G.D. equals 1.55 cubic feet per second
1 foot head equals 0.433 psi
1 psi equals 2.31 feet of head
Area of a circle equals 3.14 times the radius squared
Area of a circle equals .785 times the diameter squared
Circumference if a circle equals 3.14 times the diameter
Volume of a cylinder or pipe equals 3.14 times the radius squared times the length (or height)
1 H.P. equals .746 KWH
H.P. equals gpm times head in feet divided by 3960
B.H.P. equals H.P. divided by pump efficiency
KWH per 1,000 gallons pumped equals head in feet times 0.00315, divided by the pump efficiency times the motor efficiency
1 mg/L equals 8.34 pounds per million gallons
Dosage in mg/L equals pounds of chemical per day times 0.12 divided by M.G.D.

# INDEX

## —T—



## —U—



## —V—



## —W—



## —Z—